Second Opinions

VIKING

75 years

ALSO BY JEROME GROOPMAN, M.D.

The Measure of Our Days

Second Opinions

STORIES OF INTUITION AND CHOICE
IN THE CHANGING WORLD OF MEDICINE

JEROME GROOPMAN, M.D.

VIKING

VIKING
Published by the Penguin Group
Penguin Putnam Inc., 375 Hudson Street,
New York, New York 10014, U.S.A.
Penguin Books Ltd, 27 Wrights Lane, London W8 5TZ, England
Penguin Books Australia Ltd, Ringwood, Victoria, Australia
Penguin Books Canada Ltd, 10 Alcorn Avenue,
Toronto, Ontario, Canada M4V 3B2
Penguin Books (N.Z.) Ltd, 182–190 Wairau Road,
Auckland 10, New Zealand

Penguin Books Ltd, Registered Offices:
Harmondsworth, Middlesex, England

First published in 2000 by Viking Penguin,
a member of Penguin Putnam Inc.

3 5 7 9 10 8 6 4 2

"Decoding Destiny" first appeared in *The New Yorker*.

LIBRARY OF CONGRESS CATALOGING IN PUBLICATION DATA
Groopman, Jerome, M.D.
Second opinions : stories of intuition and choice in the changing
world of medicine / Jerome Groopman.
p.   cm.
ISBN 0-670-88801-X
1. Medicine, Popular—Miscellanea.   2. Medicine Case
studies.   3. Medicine—Decision making.   I. Title
RC82.G76   2000
610—dc21   99-36692

This book is printed on acid-free paper.
∞

Printed in the United States of America
Set in Goudy
Designed by Betty Lew

For my wife, Pam,
whose love and wisdom
bless my life

Contents

Second Opinions

Prologue

My first experience as a patient proved as instructive as all my classes in medical school.

In the autumn of 1979, while training for the Boston Marathon, I developed a nagging ache in my right hip. I assumed I had bursitis, an inflammation around the joint. My self-diagnosis proved drastically wrong.

A sports medicine doctor was the first physician I consulted when the ache slowed my pace. He gave me Indocin, a strong anti-inflammatory medication, for the presumed bursitis. "Take it easy for about two weeks," he said.

I did not want to rest. I would lose my edge and endurance. "What about a rowing machine?"

The doctor looked at me knowingly and said light crewing was probably okay.

I set the metal oars at high resistance and pulled hard, warming up my arm and leg muscles. Then I rowed with my legs extended, to maximize the effort. The familiar dull ache grew in my right hip. I ignored it. A few minutes later, a viselike spasm exploded in my lower back. Electric shocks raced down my legs. I fell to the floor.

It took many hours, lying in a fetal position, until the pain eased.

There were still tingles of electricity that played over my buttocks and thighs as I hobbled to bed.

In retrospect, the ache in my hip was not bursitis but referred pain from a nerve pinched by a bulging lumbar disc.

My wife, Pam, then a resident at the Massachusetts General Hospital, came home. Her advice, later echoed by the sports orthopedist, was that the best approach was strict bed rest, continued anti-inflammatory medication, and tincture of time.

But I wanted an immediate remedy and stubbornly believed I knew what was best. After all, my medical training had been as a student at Columbia, an intern and resident at the Massachusetts General Hospital, and a fellow at UCLA. Waiting patiently for nature to heal me seemed passive and paltry, so I doctor-shopped, seeking the second opinion I wanted to hear. I found a neurosurgeon willing to perform a limited operation on the bulging disc. The surgery did not fully return me to my prior state. There was still a dull ache in my back and hip. The marathon came and went.

In June of 1980, I left Boston for Los Angeles to join the UCLA faculty as a specialist in blood diseases and cancer medicine. Pam began her last year of medical residency there. I had not given up the idea of marathons.

One morning, after coffee at a friend's house in West Los Angeles, I stood up from my chair and abruptly collapsed. Again a powerful spasm gripped my lower back and electric shocks sped down my legs.

X rays showed no clear cause for the relapse. There were no bulging discs. I saw many consultants: rheumatologists, neurosurgeons, sports medicine doctors. Each told me that the lumbar spine is a "black box" and best left alone to heal itself.

I was emotionally frayed and bitterly frustrated by the lack of answers. The cause of my problem had to be defined and aggressive solutions applied. I was determined to be permanently repaired.

"You'll be up and running within two weeks," a burly orthopedist in private practice in Beverly Hills told me cheerfully.

He asserted that I had "instability" of my lower spine. A fusion, done by harvesting bone from my pelvis and inserting it along the ridges of my lower spine, would create an internal brace and fully restore my mobility. His partner, a neurosurgeon, wasn't so sanguine, but I was not deterred. The heady promise of the orthopedist made moot any other consideration.

I awoke from the surgery in the intensive care unit. My lower back felt woody and numb, and at first I thought I was cured. But moving my legs, or even just flexing my toes, triggered waves of such pain that my previous symptoms seemed minor.

I had hemorrhaged during the operation. The surgeons were unsure why. Perhaps the chronic anti-inflammatory medication had weakened my clotting system, although everything seemed in order preoperatively.

The agony did not relent. I was told I had neuritis, that my spinal nerves were irritated from the spilled blood and resultant scarring. The orthopedist suggested he operate again to free the nerves. I was confused and uncertain. At Pam's insistence, I declined.

For three months I lay on ice to numb the stabbing pains. I was given strong analgesics, Percodan and other narcotics. They made me nauseated and dopey. I could not focus and think. Books were a blur. I had little to say in conversation. I was despondent and became terrified that this would go on for the rest of my life.

I finally realized that my desperate belief in a perfect solution was a fantasy. I also realized that it was up to me, in part, to try to rebuild myself. I consulted a specialist in rehabilitation medicine. The narcotics were discontinued as I began physical therapy. The first sessions were awful, my frozen legs resisting even minor passive movements. I gradually extended my stride by supporting myself on parallel bars submerged in a hot pool. It took nearly a year for me to walk more than a few yards.

I returned to my career as a hematologist and oncologist at UCLA. A cot was set up in my laboratory, so that I could rest regularly. After I made rounds with the residents and students, I often had to lie on the floor of the conference room and discuss my patients while supine.

I have never fully recovered from the surgery. Not a day passes when I don't fail to think of my headstrong decision, because of the limits on my functioning. The pace and length of my stride are tightly constrained. Soft seats, like those on an airplane, fail to support my lumbar spine, and after a few minutes in them, my back goes into spasm; I carry a customized back support when I travel or go to a meeting. If I bend incorrectly, or quickly lift my heavy briefcase, or overtax muscles by standing too long on rounds, I suffer a siege of back and leg pain.

From this debacle and my chronic debility I developed a more tempered view of medical interventions and an abiding sense of humility about my profession and my own practice.

When patients and I sit in the quiet of my office and consider which options to choose, I often recount my experience. It brings me closer to them, knowing that I was on the examining table, swept up in the same tempest of confusion, fear, and frustration, vulnerable to all sorts of advice.

I make no pretense of omniscience. Decisions about diagnosis and treatment are complex. There are dark corners to every clinical situation. Knowledge in medicine is imperfect. No diagnostic test is flawless. No drug is without side effects, expected or idiosyncratic. No prognosis is fully predictable.

Still, there are important landmarks that help doctor and patient successfully navigate this uncertain terrain. A clinical compass is built not only from the doctor's medical knowledge but also from joining his intuition with that of his patient. This melding of minds occurs when the physician probes not only his patient's body but also his spirit, considering not only the physical repair required but also the psychological and emotional needs. Eliciting a patient's intuitive sense of his condition is not simple. It takes time and open dialogue to build trust with a person and to encourage him to express himself.

I wrote this book to provide an inside view of the complex and rapidly changing world of medicine and thereby help people make informed decisions. While there are no pat guidelines applicable to every patient and every situation, the chapters that follow retell critical moments that have forever shaped my thinking and practice—not only for my patients but for my family and myself. It is these times—both when my opinions and actions proved right and when I seriously erred—that I have sought to portray in the clinical dramas which occurred around me.

One such medical encounter occurred with Karen Belz, a longtime friend. Karen's family was plagued by breast cancer—her mother had died from it, and her sister, Ruth, was undergoing chemotherapy for widespread metastases. Karen was struggling with the question of being tested for a mutation in the breast cancer susceptibility gene BRCA, which is frequent in people of Eastern European Jewish descent. Did Karen really want to know, and what would she do, if she tested positive?

I came home from the hospital that evening and discussed Karen's dilemma with Pam, now a specialist in endocrinology and metabolism. I assumed she would agree with my clear logic and readily endorse my recommendation of prophylactic surgery to Karen. I was stunned when Pam took a very contrary point of view. What I took as the optimal choices were seen as problematic by a doctor whose expertise I admired and whose values I shared. What had seemed clear-cut was not.

My experience with Karen Belz and her treatment appeared as a story in *The New Yorker* magazine several months later and is included here. Her situation emphasizes that clinical decisions cannot be formulaic but must fit a person's beliefs, background, and desires. The risks we are willing to take, and what we are willing to give up to try to live, differ for each individual.

The other seven stories are published here for the first time. Each was chosen to illustrate a particular dimension of medical decision making.

Some choices are urgent and the consequences immediate. In the ER or operating room, decisions can spell life or death within minutes. My firstborn son, at ten months of age, developed an intestinal obstruction and almost died due to physician misjudgment. His story opens the book. Other chapters highlight seemingly minor decisions, made in less urgent circumstances, which proved to importantly determine the health and welfare of a person—their effects not manifest for months to years. I also recount instances when I had no clear understanding of what was happening clinically. Nonetheless, choices of great risk had to be made.

In a predictable world, clinical decision making would be a well-defined, scientific exercise with set methods for diagnosis and treatment. One increasingly vocal health care camp, populated by technocrats and managed care administrators, promotes this view. They believe that "decision trees," with branching algorithms and formulae, are necessary and sufficient for the practice of "effective medicine." The unfortunate truth is that this is not possible. People adapt differently, physically and emotionally, to each illness and react in varying ways to a given therapy. This means that diagnosis and treatment cannot be strictly bound by generic recipes, but must be made individual, to be consistent with the particular clinical and psychological characteristics of the person.

It is imperative that a physician has deep knowledge of his patient and his disease and ready access to first-rate technology. This conclusion comes from one of my most painful family experiences. In the spring of 1974, while still a student at Columbia, I was awakened in the middle of the night by a telephone call from my mother. My father, the person whom I loved and looked up to most in the world, had had a heart attack. His doctor was a general practitioner associated with a small community hospital in Queens. The ambulance was directed to that hospital rather than another nearby, larger center. I rushed from my dormitory in Manhattan and arrived at his bedside. No cardiologist was available. There was no intensive care unit in that hospital. The treatment my father received was rudimentary, and he died.

I reflected on that moment at each subsequent step of my clinical training—at Columbia, at the Massachusetts General Hospital, at UCLA. In these institutions, I saw what care could be. Other men in their fifties were saved by rapid and expert intervention. Advanced technologies, like aortic balloon pumps, angioplasty, and respirators, relieved the failing heart, opened blocked coronary arteries, and supported the fragile system. Of course, there were times when the best cardiologists and the best treatments failed. But at such moments, the doctor could assure the family that everything that could be done was done.

I committed myself to practicing medicine with the rigor and expertise my father had not received and to steering patients to the best medical centers that offered every chance for help.

But what is the "best" hospital or the "best" doctor?

When my grandfather had Alzheimer's disease, he was referred to a specialist touted as at the top in his field, based at a prestigious medical center. This distinguished doctor proved far from the best for the emotional and logistical needs of my grandfather and our family.

Thus, providing outstanding medical care demands not only scientific skill but also an open heart. These days, yet another attribute is needed. The clinical landscape is changing rapidly, much of the upheaval due to HMOs and managed care. This new world of medicine is populated by overseers who dictate your options and have interests that may conflict with your own. So you need more than a caring physician competent in physiology and pathology and pharmacology—you need someone who knows how to bypass the obstacles of bureaucracy and business, to maneuver in the system and get things done for his patients.

It is my hope that the stories in this book will enable the reader to make medical choices—his own, or for family or friends—with greater knowledge and deeper intuition, thereby contributing to the remedy of illness and enjoyment of health.

Jerome Groopman
Brookline, Massachusetts
May 1999

Our Firstborn Son

On the morning of July 4, 1983, Steven, our first child and then nine months old, was crying bitterly and refusing to nurse. Usually good-tempered and not prone to colic, he seemed inconsolable. Something was wrong.

It was the last leg of our journey back to Boston after three years working at the University of California. We had flown nonstop from Los Angeles to New York the prior evening and then driven to Pam's parents' house in Connecticut to spend the night. Steve had missed his regular nap on the flight and was too cranky to rest in the car.

We were first-time parents, and as physicians, Pam and I vowed when Steve arrived that we would not diagnose or treat our own children. It was not only the concern that our clinical judgment would be clouded by parental emotion. We were also relatively ignorant about pediatrics. Our knowledge of infant health and disease came from a brief introductory course during the third year of medical school. And living in California, far from our families, we had no one to mentor us in the common sense of baby care. So we studiously read how-to books by popular pediatrician writers and checked Steve's milestones in growth and development against their charts. He had been a healthy infant.

It had been an animated homecoming that hot July evening in Connecticut. Steve was my in-laws' first grandchild, and they were delighted that we had returned. We had talked together around the kitchen table late into the night, shamelessly eating Häagen-Dazs and homemade cookies.

Pam and I assumed Steve would sleep late the next morning, given the three-hour time difference and his missed nap. But the night-table clock read just 5:12 when we heard his plaintive cries.

"You check him," Pam muttered as she fixed the blanket over her head like a cowl in a determined effort to return to sleep. We attended to our son pretty much according to our biological clocks. I am a morning person, Pam a night owl. Before 3 A.M., she gets up; after that, it's my turn.

"Did he nurse last night?" I asked, wondering if Steve was hungry.

"Not much. He started, and then stopped, so I just put him down to sleep."

I navigated through the clutter of luggage, laundry, and shoes on the bedroom floor to Steve's Portacrib. I noticed he was lying on his side rather than in his usual position on his back. His broad silky crown was pressed into the crib's foam bumpers, obscuring his face.

"Hey, it's too early to get up, Zalman Leyb," I said, using his Yiddish name he was given in conjunction with his English one. In the Ashkenazi tradition, children carry the Hebrew or Yiddish names of deceased relatives, to perpetuate their memory. Zalman was my father's name and is the equivalent of Solomon. Leyb was Pam's grandfather's name and means "heart" in Hebrew and "lion" in Russian-Yiddish. We had chosen these names not only as a bridge to the past but as a hope for the future that our firstborn son would be blessed with a heart of wisdom, courage, and compassion.

A sharp sulfurous odor assaulted my nostrils. It rose from a heavy black stream that was lazily meandering down Steve's fleshy thighs.

"Must be a whopper," I said. I held my breath and quickly unhitched the adhesive waistband of his Huggies. On Steve's bottom was a wide swath of mucoid diarrhea. I noticed speckles of maroon dotting a tarry gelatinous stool.

"His poop is strange," I called to Pam.

Pam threw the blanket from her head and unsteadily made her way to the crib. I offered the open diaper for her inspection. Her brow tensed and cheeks stiffened.

"It looks like melena," she said gravely.

Melena is the medical term for blood in the stool. Blood, when it is digested in the intestines, gives stool a distinctive tarry appearance and pungent odor.

"And he feels a little warm to me."

I placed my palm next to Pam's on Steve's flushed forehead. Not more than a 101, I estimated.

Pam stood next to the crib, warily observing the baby.

"I'll clean him," I said.

I volunteered for such chores, knowing that my father never once, according to family lore, changed a diaper or gave a bottle. I aimed to be the modern husband-partner. But my initiative at that moment was different: I wanted to be occupied in a menial task that suspended thinking. Melena indicated internal bleeding, and although it could be caused by gastritis, it also might be the first sign of something more serious, like a tumor.

As I wiped away the viscous stool, an angry rash emerged. I liberally sprinkled Desenex powder over Steve's crimson buttocks and then snugly enclosed him in a new diaper.

"Let's see if he's hungry," Pam suggested, deftly lifting Steve from the crib and holding him securely across her chest. She began to coo softly and to stroke his back. Steve faced me, and I noted again how he had inherited Pam's unusual eyes, the left one sea green, the right one lapis blue.

Pam settled into a large wicker rocker set near the bed and unsnapped her nursing bra. Her swollen breast was directed to Steve's thin closed mouth. Pam lightly stroked his dry lips with her offering. Beads of ivory milk coalesced into a halo around her areola. Steve responded quickly, sucking greedily, but after less than a minute tensed his cheeks, whimpered, and released the nipple. The milk streamed down his chin.

"What's the matter, sweetness?" Pam asked gently.

Steve emitted a harsh grunt and then flexed his legs toward his belly. He repeated this brusque motion several times, breathing noisily.

"Here, let's try again."

But to no avail. Each attempt to nurse was met with aborted gulps followed by sharp jerking of his legs.

We laid Steve on his back in the crib, but he fussed until he turned himself on his side with his knees wedged toward his belly.

Pam balanced on her toes next to the crib, her hands whitening in clenched fists as she stared down at the baby. She was poised like a lion that had picked up the scent of danger and was ready to defend her young.

"Where do we find a pediatrician on July Fourth?" she tensely asked.

It was not obvious, and we began to think out loud. We first would look for neighbors with children and ask them to call their doctor. The doctor might be away, or reluctant to respond since we were not his patients. We'd then be forced to go to a hospital emergency room. Norwalk and Bridgeport were equidistant from Pam's parents. Emergency rooms are notoriously busy on holiday weekends. Despite our suspicion of melena, Steve was not frankly bleeding, and would not be judged an urgent case. We likely would sit for hours, triaged to the back of a queue of people in greater need, with heart attack, trauma, sepsis. There also would be a single primary care doctor covering such a local ER, without a pediatrician in attendance. Yale–New Haven Hospital, a larger teaching institution, would have a pediatric intern on site, but it was nearly two hours away. And we both knew the inside joke among physicians: Never get sick in July. The new interns begin then. Some few days ago they were bewildered students, but on the first of that month they are instantly transformed into decision-making doctors by donning a white coat with the monogram MD.

Pam always stocked liquid Tylenol in Steve's travel bag. A dropperful would do no harm, we agreed. Lowering his fever might improve his disposition and facilitate his nursing. The Tylenol at first stayed down, but as we lay Steve back in the crib, he grimaced, abruptly drew his knees to his chest, and vomited. The cherry color of the medicine grew into a sickly brown as it diffused in a pool of bile on the sheet.

The Clarks, young neighbors who had purchased their home through my mother-in-law's real estate agency, called their pediatrician, Dr. John Burgess, at home. We explained our dilemma to him, and without a hint of hesitation, he instructed us to meet him at his office in fifteen minutes.

Steve had become calmer after vomiting, and the brusque flexing of his legs ceased. We easily found a parking spot on the tree-lined street in front of the town's medical building. Dr. Burgess was waiting at the entrance and greeted us with a polite smile. He was a trim man in his early sixties, with short-cropped white hair and half-glasses that rested low on his nose. Despite the steamy summer heat and the holiday, Dr. Burgess wore a starched button-down blue shirt, paisley bow tie, and knee-length white coat with his name embroidered in blue script over the left breast pocket. His crisp professional appearance reassured me. In Los Angeles, I was disturbed by the indifferent attire of many of the younger doctors: the unkempt hair, running shoes, and jeans bordered on sloppiness and did not indicate to the patient the sense of order and attention to detail essential to diagnosis and treatment.

Steve submitted docilely to the pediatrician's experienced prodding and auscultation. After a few minutes, Dr. Burgess nodded to us that his exam was finished. Pam picked up Steve and dressed him in a fresh diaper and clothes.

"Just an intestinal virus," Dr. Burgess announced with confidence. "Keep up the Tylenol—"

"But it had the look and smell of melena," Pam interjected.

"It wasn't," Dr. Burgess quickly retorted, his eyes holding Pam's. "I was once an overanxious doctor-parent, too." He paused and his face softened with a knowing grin. "My brood is grown up and long out of the house. But you never stop worrying about children. And a little knowledge is a dangerous thing. He'll be himself in a day or two. You have a basically healthy baby."

A wave of warm relief slowly passed over me. I looked to Pam. She maintained a stony silence.

"Like I started to say, continue the Tylenol if he feels feverish, every four to six hours, for the rest of the day. And just let the diarrhea flow. Nature knows how to get rid of the bad humors. Keep him hydrated today with warm sugar water from a bottle. It goes down easier than breast milk with an intestinal virus. You can try to nurse him again tonight. And don't bundle him up on the ride back to Boston. It's going to reach into the nineties today. You'll overheat him." Dr. Burgess smiled again, removed his white coat, and donned a blue-striped seersucker sports jacket. He escorted us out to the front door.

"Sorry to have bothered you on the holiday," I repeated as I shook Dr. Burgess's hand good-bye.

"My pleasure to assist."

I was filled with that profound gratitude of the patient for the doctor, for not only selflessly attending to a loved one but also for taking away my fear. I carry a special affection and admiration for pediatricians, realizing early in my training that I could not be one. It was too emotionally wrenching for me to attend to very sick children. During my brief medical school tour through the pediatric wards, I cared for children with ultimately fatal problems, like cystic fibrosis and leukemia and malformed hearts, and ended each day profoundly upset, pained that such young and delicate innocents were suffering. These feelings overran my thoughts on rounds, and I couldn't focus on the clinical issues of the cases being presented.

Since Steve's birth, I had tried to subsume much of the anxiety that comes with parenting in general and my past exposure to seriously ill children in particular. I am not a hypochondriac, but as a specialist dealing daily with life-threatening blood diseases and cancers, I found myself imagining these maladies striking my son. Similarly morbid thoughts seized my mind at unexpected moments. That first year of Steven's life, at the Passover seder, the ritual recitation of the Exodus from Egypt, I had shuddered when Pharaoh orders all male Hebrew children thrown into the Nile. I was especially gripped by the Tenth Plague, when the Angel of Death passes over all the firstborn of Egypt, man and beast, and kills them in retribution. These were chilling illustrations of the shattering power of the loss of a child. Only the Tenth Plague could break the obdurate Pharaoh, all his authority and possessions made meaningless in its wake.

Dr. Burgess's words reminded me how easy it was, with a little knowledge, to allow my fears to race ahead of rational thinking. As a first-time father with a "basically healthy baby," I remarked to myself, *Better to be a layman, in the bliss of ignorance, than a neurotic doctor.*

"It's already one o'clock," I said to Pam as we exited the medical office building. "Let's pack and try to get up to Boston before dark."

Pam looked hard into my eyes. "I hope he's right."

Before we left Pam's parents, Steve kept down the full six-ounce bottle of sugar water recommended by Dr. Burgess. The July sun was high in the sky, and the rental car had no air-conditioning. We drove with all windows open. The warm breeze seemed to soothe

Steve. Traffic was light and we enjoyed a steady cruising speed on the Merritt Parkway. Most people had left early for beaches or parks to escape the heat and claim a good vantage for the evening fireworks sponsored by the local towns. By the time we approached the exit to Interstate 91 North, Pam had fallen asleep in the back beside Steve. He sat awake, docile in his car seat.

I relaxed and began to think of our new life back in Boston. It was a career risk to return. We had just begun to establish ourselves at UCLA. I had succeeded in setting up my own laboratory, secured a five-year grant to study blood cell development, hired an adept technician, and was publishing scientific papers. The dean of the UCLA Medical School had prevailed upon me not to leave. The ticket to success in academia was to investigate a problem "an inch wide and a mile deep." AIDS seemed to be such a problem. It had just appeared, the first cases reported by our group in Los Angeles and colleagues in New York. This was an unparalleled opportunity: rarely in a doctor's career does he have the chance to describe the manifestations of an entirely new disease, contribute to identifying its cause, and work to create effective therapies. There is no AIDS in Boston, the dean had said; you have funding, lab space, and a growing reputation. You're throwing away a golden chance to make your career on this new rare syndrome.

But I always had a contrarian streak. When I was a youngster, my father gave me the nickname "Upside-Down Jerry," because I stubbornly insisted on trying to do things against given directions. Authority was heeded only if it seemed sensible, not because it was authority per se. Now a group was moving from the Massachusetts General Hospital, where both Pam and I had trained, to create a new department of medicine at a small neighboring institution, the Deaconess. The Deaconess was excellent in clinical care but lacked the research depth that marked the other Harvard teaching hospitals. I could build a new research enterprise in blood diseases, can-

cer, and immunology from the ground up. Risk and change ener-
gized me.

Pam had wanted a change, too, in the form of a flexible year with
the new baby. She had been on track since college, engaged full-
time in her work. Now she planned to break away from that path.
She had taken a part-time job, teaching interns and residents at the
Waltham Hospital, a community-based center associated with Harvard's Brigham Hospital. With the position came the perk of a free
house in that blue-collar town some fifteen miles west of Boston. A
few months earlier, when we came to Waltham to finalize our
arrangements, we had inspected the two-bedroom prefab with its
single bathroom and backyard abutting the parking lot of the Mc-
Namara Concrete Company. From the kitchen window you saw
scores of steel gray concrete mixers standing like soldiers in forma-
tion. In California, our home was on a cliff overlooking the Pacific
Ocean with a view of Catalina. "Are you sure you are ready for this?"
Pam had pointedly asked. I had paused, and then said I was ready for
anything.

Shortly after we connected to Interstate 84 in eastern Connecticut, Steve began to fidget and whimper. Pam snapped awake at the
sound of his cries. The car radio said the temperature was 93 and the
humidity 87 percent. Pam felt his cheek. He was warm and likely
thirsty. We stopped at the next highway rest area and settled under
the speckled shade of a motionless willow. Pam sat cross-legged and
cradled Steve, speaking to him encouragingly. But he continued to
fuss and would not take any sugar water from the bottle. I slid my
index finger into the bottom of the new diaper he was given after Dr.
Burgess's examination. It was completely dry.

"He hasn't peed at all."

Pam's face tensed and her eyes focused on Steve's squirming legs.

"Let's sponge him with some cool water from the fountain and get
to Boston," I suggested.

Pam silently indicated her assent.

We drove quickly up Interstate 84 to the Massachusetts Turnpike and headed east. I knew the exits on the route to Boston by heart: Palmer, Auburn, Worcester, Framingham. As we passed each, I noted the mileage on the odometer and estimated the distance to the next one. Pam again tried to coax Steve with the bottle of sugar water but without success. It was too early to give him more Tylenol, and in line with Dr. Burgess's instructions, she did not try breast-feeding.

As we passed Framingham, the penultimate exit, Steve tried to flex his legs upward but was restrained by the straps of the car seat. Each aborted attempt ended in a piercing cry, like a kitten stepped on by a heavy boot. Pam tried to comfort him, stroking his pale and dry forehead.

I began to sing "Humpty Dumpty," which we played from a music box for Steve at bedtime, hoping the familiar hypnotic rhythm would be soothing. But my voice wavered as the meaning of the words took on a dark resonance:

> *Humpty Dumpty sat on a wall,*
> *Humpty Dumpty had a great fall;*
> *All the king's horses*
> *And all the king's men*
> *Couldn't put Humpty Dumpty together again.*

I stopped and tried to think of another song we sang. "Jack and Jill" ended in a broken crown and tumbling down a hill. "Rock-a-Bye-Baby" climaxed in a free fall of cradle and child. I remained silent.

"Try 'Hot Cross Buns,'" Pam offered.

I tried to remember the verses, but the song became dispersed in the waves of fear breaking in my mind. It *was* melena. Steve *is* bleeding internally. From an intestinal *tumor*.

My hands became unsteady and I worried I'd miss the turn off the Mass Pike onto the twisting surface roads into Waltham. I calmed myself by repeating Dr. Burgess's injunction, to try not to be a physician-parent with limited knowledge who easily panicked.

"We'll settle in at the house, cool him down in the bath, and then see how he is," I said, talking as much to myself as to Pam.

We pulled into the driveway at dusk, the last tired rays of the summer sun casting a pall over the small house. Our new neighbors were in their front yard, drinking cans of soda and beer next to a smoldering barbecue. I stopped to wave to them and then began unloading our bags. Pam freed Steve from the car seat and quickly carried him into the house.

The bags felt extraordinarily heavy, and I realized I was exhausted. I had slept less than four hours the night before, and except for the respite following the visit to Dr. Burgess, had been in a sustained state of anxiety since. My infant son was in escalating pain. If this were an intestinal virus, as Dr. Burgess asserted, it was wreaking havoc inside his tiny abdomen.

I looked to Pam for reassurance. I admired her as a highly competent physician whose attention to detail and clinical judgment were formidable. At the hospital, she functioned in a deliberate manner, coolheaded and in control, even in the most harrowing emergencies. But her reassuring mien was gone. Her eyes had a furtive look and her voice was unsteady when she asked if everything had been unloaded from the car.

"Let's give it an hour or two," I said as I glanced at my watch and noted it was just after eight. "Maybe if he's out of that hot box of a car and in your arms he'll relax and take the sugar water again."

Pam warily agreed as she searched for a kettle in the kitchen cabinets to warm the water. I busied myself upstairs in the bedroom by putting away clothes.

I followed the order I'd used since first leaving home for college: socks and underwear in the upper drawer; T-shirts and shorts below;

with heavier items, jeans and folded dress shirts, at the bottom. Then I arranged my pens as I liked them on the night table. These set, I began to organize my books on the shelf, the upper shelf for medical texts, the lower for literature. As I put away my belongings, I strained to hear noises from downstairs, and when no cries were apparent, told myself that in the more comfortable setting of the house Steve would settle down. The stomach bug would run its course, as Dr. Burgess had said. Steve was a basically healthy baby.

"Jer, come down! Come down!"

I dropped a pile of books and bolted from the bedroom. I took the steps two at a time, the rickety banister barely supporting my weight. Pam was kneeling on the beige carpet. Steve was on his back next to her, his face ashen and his breathing forced in short gasps. He was desperately flexing his knees to his chest, his arms flailing at his sides. He looked like a wounded beetle in the throes of dying.

"Let's take him to Children's Hospital—now!"

I responded to Pam's command by grabbing the ring of keys I had left on the table in the foyer. Pam struggled to hold Steve securely as we ran to the car.

I retraced our route back down Highland Street to the Mass Pike. On the highway, I forced myself to keep a steady 55; this was no time to be stopped by a cop or lose control and hit another car. Pam sat with Steve pressed to her chest. He was having spasms of pain every few minutes. As each came on, he would again draw up his knees and shriek. Pam stroked his back, softly repeating, "It's okay, it's okay."

I again focused on landmarks to mark our progress: the Newton Centre sign indicating we were halfway there, then the Allston–Cambridge toll, the reverse curve of Storrow Drive at Boston University, the Fenway overpass at Kenmore Square, and finally Longwood Avenue, the heart of the Harvard Medical area. The car clock read 8:22 when we left Waltham. It took twenty-one minutes to arrive at the Children's Hospital.

I spied three empty parking spots near the hospital entrance and pulled into one, wondering if it was a sign that the ER was not busy.

We hurried through the sliding automatic doors and then halted in the ER waiting room. It was a tumultuous scene. Every chair was taken. Many parents stood with squirming infants in their arms. Toddlers were sprawled on the floor.

The waiting room, with all its chaos, was unexpectedly comforting. We were finally in a hospital, and not just any hospital but Boston Children's, one of the premier pediatric institutions in the world. And I knew the place well. I had worked for a year at Children's in a research laboratory in the department of a prominent hematologist, Dr. David Nathan. A burly, avuncular man, David relished the role of "godfather" to several generations of hematologists whom he trained at Children's and then placed in academic positions throughout the country. When Pam would work nights as a resident at Massachusetts General Hospital, I would conduct experiments late into the evening. I often went for a midnight snack in the hospital cafeteria before taking the last trolley home. The route from my lab to the cafeteria required passing through the ER waiting room.

I located the triage desk and moved toward it with a determined stride. Pam, gripping Steve, followed in my wake.

The triage nurse was a wizened, plump woman whose name badge read "McArdle," with no first name. She quickly surveyed us with her trained eyes, flipped the intake pad to a fresh page, and briskly obtained the essential information: names, address, home phone, employment, presenting problem. I couldn't remember the phone number at the new house and fumbled through my wallet for my insurance card until I realized it was for the UCLA clinics.

"We'll deal with that later," McArdle said as she put the pad aside and directed us to the swinging doors behind her that led to the ER proper.

"First door on your right. You'll be seen shortly."

The room was a thin rectangle that accommodated two chrome chairs with black vinyl seats, a short examining table, and a row of instruments on the wall: blood pressure cuff, ophthalmoscope, otoscope, suction apparatus. I sat on the chair next to the examining table, noting how unyielding it was. A painful knot grew in my lower back. I shifted in the chair and tried to ignore it. Pam would not sit but paced the room clutching Steve to her chest.

After a few minutes, an ER doctor walked in. He plucked McArdle's triage form from a Lucite box on the open door. In his late twenties, he was strong featured and wore the V-neck blue scrub shirt of a surgeon. Tufts of wiry brown chest hair formed a nest upon which his stethoscope rested. The stubble on his cheeks suggested he hadn't shaved that day. I saw his hospital badge and read, "Scott Warren, MD."

"Dr. Warren, I'm Dr. Jerry Groopman, and this is my wife, Dr. Pam Hartzband," I said in as welcoming and deferential a tone as I could muster.

"Uh-huh," he responded, not picking his head up from the telegraphed notes on the sheet. After fully reading the form, he introduced himself.

"I'm the surgical moonlighter, a PGY3 covering the ER."

I took some comfort from this, that he was in his third postgraduate year of surgical training. Abdominal pain is properly first triaged to surgeons in the ER.

I was tempted to ask him if there was a senior attending surgeon available, but refrained. I had been a resident and now was a teacher of residents. Patients and families who immediately challenge a resident's authority by demanding the attending doctor may alienate someone who can be an important ally. Senior residents at major university hospitals often are the best and the brightest in medicine—young, enthusiastic, hardworking, fresh from the latest teaching. True, they lack seasoned experience, but they also are more hands-on, and often perform better in the ER than older staff, because they work there day in and day out. Of more concern was that some residents become perverse when they encounter patients who are difficult or noncompliant. At Mass General during my training, such patients were given a denigrating acronym by the interns and residents: GOMER, for "get out of my emergency room." The vitriol embodied in the term sometimes impaired sound judgment or led to neglect of serious complaints. Dr. Warren was the most important person in our universe at that moment, and I was eager to win him to our side.

"Are you any relation to Chip Warren of the distinguished Warren medical family? Chip was in Pam's class at Harvard Medical School and one of my students when I was a junior resident at the Mass General."

This was not idle social conversation. My words were meant to say: *We are both doctors, Harvard trained. Our past is like your present. We may even know people you know. Please treat our son with special care and concern.*

"No, no relation. A lot of attendings ask if I'm part of the Warren dynasty. But I'm not. Now, when did your son first seem sick?"

He directed the question to Pam, who responded with a chronicle of the events of the last thirty-six hours. It was Pam at her best, the organized thinker with flawless recall for detail. She began by stating that Steve had been previously healthy, the product of a normal delivery who had achieved all his developmental milestones. She then described the missed nap on the flight from L.A., his cranky awakening the next morning, and emphasized the color, viscosity, and odor of the morning stool that she thought was melena. Her recounting of Dr. Burgess's evaluation and contrary opinion was objective, without editorial comment. Pam then summarized Steve's temporary respite after the Tylenol and sugar water, his renewed whimpering on the trip up, the stop at the highway rest area in Connecticut, his dry diaper and refusal to drink, which meant that he was without any liquid or nourishment for the past eight hours. She ended with his gasping breathing punctuated by shrill cries during the spasms of pain that brought us to the ER. I sat silently, ready to comment on an omitted event or observation, but the clinical history was complete.

"Has his diaper been dry? How many hours since he last had fluids?" Dr. Warren asked.

I saw Pam's face register first uncertainty and then concern. A sick taste welled in my mouth. I locked on to Pam's eyes and spoke without words: *You already told him that; he's not listening carefully; he wasn't paying close attention.*

"What d'you got in here, Scott? What is it? A good case?"

I looked at the door and a grinning pie-faced young doctor stood under the lintel. Also wearing a blue surgical scrub shirt, the hip pocket of his short white coat bulged with the *Washington Manual*, a primer that interns carry to reference diagnoses and treatments on the run.

I met the intern's gleeful eyes with mine. I felt my hands tremble, my jaw tighten, my chest heave. A thunderous rage rose up from the deepest recesses of my being.

"Who the hell are you?" I bellowed. "Who the hell are you? My son is not a 'good case'! My son is not an 'it'!"

I felt I might kill the intern, literally grab his neck and snap his spine with the force that was exploding in my limbs. In my mind's eye, I saw myself pounding his flopping head against the pale green wall of the ER exam room.

The intern looked at me in shock and then turned perplexed to Dr. Warren.

"Get out of the room! Now! Now!" I yelled.

Dr. Warren nodded to indicate that his intern should leave.

I stood paralyzed for a long moment and then retreated to the corner of the room, lowering myself onto the chrome chair. My blood was coursing in forceful pulses through my head, my breathing still sharp and fast. I willed my fists to release, my arms to rest at my sides. I reached for the handkerchief in my pocket and wiped the lines of sweat collecting at my brow.

"I'm sorry," I finally said, my eyes fixed on the floor before me. "I know that is how we doctors sometimes talk about patients. But he's our child." I raised my head to speak directly to Dr. Warren. "We're just scared out of our minds."

Dr. Warren nodded to accept my apology and then politely asked if he could examine the baby.

Pam and I hovered close by, observing the resident's every move. Steve had tired, his efforts in drawing his legs to his chest less forceful than earlier in the evening. I glanced at the large Seth Thomas clock on the wall. Its numbers were in bold black, and it had a silent, sweeping red second hand. It was already 9:35.

Dr. Warren worked silently, lingering with his stethoscope over each quadrant of Steve's belly. Finally, he seemed satisfied with what he heard and slung the stethoscope around his shoulders like a mantle. Pam reached over and gathered Steve back in her arms.

"Rushes of bowel sounds, then quiet. Very classic."

"Intussusception?" Pam asked, quicker than I to translate the physical findings into a diagnosis.

Dr. Warren said yes.

Intussusception is a telescoping of one segment of bowel into another, resulting in acute intestinal blockage. In infants, it often occurs without apparent cause, but can also be due to a viral infection or a tumor growing in the intestine. One segment of bowel becomes abnormally heavy as the tumor grows in it or viral infection inflames it. This weighted area acts as a so-called leading point. Waves of peristalsis, the rhythmic muscular contractions that move food and stool down the intestine, push forward and meet the resistance of this weighted segment. This causes the preceding piece of intestine to telescope over the heavy segment and swallow its partner in a strangling gulp. Each peristaltic wave triggers searing pain as the lighter piece of bowel further engulfs its heavy partner and stretches the delicate intestinal nerves like subjects of torture on an inquisitor's rack. This explained the excruciating spasms of pain Steve had suffered over the last day. The blockage accounted for his bilious vomiting.

"We'll get some blood tests and then do a barium enema."

Dr. Warren explained the barium enema would confirm the diagnosis and possibly open the obstruction. The air pressure from the rectum through the enema was often enough to push back the telescoped piece of bowel. Neither Pam nor I was expert enough to know if this was the right first approach, but it seemed logical. We surrendered Steve to Dr. Warren, who transported him on a stretcher for his blood tests and then to X ray.

We walked up the flight of stairs to the empty reception area of the Radiology Department and sat in its first row of chairs. I grasped Pam's thin, moist hand and offered hollow reassurances: "It will all turn out fine. . . . We're at Children's, he's in good hands." She replied with a wan smile.

We both knew that we could not know—no one knew—how it would turn out, whether Steve would quickly recover with the simple maneuver of the barium enema, the blockage due to a virus that had inflamed his bowel, with this nightmare easily ended, or whether he required surgery, where some serious pathology would be revealed, a hidden tumor, incurable, our nightmare just beginning.

I closed my eyes to pray but couldn't focus my thoughts. A chill passed through my body.

The world seemed to become motionless and without sound, as if I had passed into a frozen vacuum.

Is this the prelude to death, my son a part of me, my soul sensing the eternal stillness of the dimension he is entering?

I felt the chill penetrate deeper into my bones.

The chill is the icy breath of the Angel of Death, searching for my first-born son. Did I forget to paint lamb's blood above our door when we left Los Angeles?

I began to speak to myself, firmly and deliberately.

Gain control of your thoughts. This is a medical emergency in Boston, not the Exodus from Egypt. There are no signs and wonders. Steven is not being punished in retribution for his father's sins, the times you were inconsiderate, ungenerous, arrogant, the occasions you turned to the idols of power, money, fame.

Pray.

God, please restore my son. If you do, I will . . .

What? Attend synagogue daily for a year? Give more to charity? Strictly observe the Sabbath?

Prayer is not bartering with God.

Then how do I pray for my son's life?

Like you prayed for your father's life. In an eerily similar moment, in that small, ill-equipped, understaffed community hospital in Queens, New York, waiting to see if he would survive. Then, too, time felt halted and the motion of the present seemed to pause as the future was determined.

Please, God, save my beloved father.

But he had died, gasping for air before my eyes. My prayers had not been answered then.

And now?

"Dr. Groopman?" The radiology resident was standing before me. "Yes?"

"It is an intussusception. But the enema did not open it. Sorry." The resident paused. "And good luck."

For a moment I was confused. Where was Steve? Who was attending to him? I slowed my growing panic and then realized that, according to standard hospital procedure, he was being returned in the special patient transport elevator to the ER, with a nurse accompanying him, and we would meet him there.

Pam and I walked hand in hand down the hospital corridor. The reflected fluorescent light cast a ghostly aura on her face. Her usually strong features, the high prominent cheekbones, piercing eyes, and calm smile, were bleached away. I again felt deeply exhausted. For a moment I feared I would collapse. I forced myself to stand erect and appear strong as I walked with her down the stairs to the ER.

"See how the dilated loops of bowel stop right here?" Dr. Warren said as he held the X-ray films up to the ceiling light.

I saw what looked like a child's unfinished chalk sketch of balloons. Their tense white upper borders stretched up against a black background. But they had no bottoms, their lower halves lost in a sea of gray.

"You can follow the loops to the air-fluid levels," Dr. Warren added, pointing on the X ray to where the balloons disappeared into the gray puddle of liquid and stool that had accumulated from the blockage.

"You know it didn't open after the enema," he continued.

I felt my eyes tighten.

"And we just got the blood work back. His white count is up—29,000—with a left shift."

I didn't know if that was typical for an intussusception, but an elevated white blood cell count and "left shift," meaning the appearance in the bloodstream of immature white cells, were often a harbinger of major infection or tissue damage.

"He seems very tenuous," Pam said. "I guess you're going to operate to relieve the obstruction."

"In my clinical experience," Dr. Warren stated with calm gravity, "we can safely observe your son overnight. It's premature to operate. And it's already past midnight."

I glanced quickly at the Seth Thomas clock: 12:17.

"I'll sign him over to the morning shift at six. They'll reassess him then. Meanwhile, he'll be kept hydrated with intravenous fluids.

You can stay in the room with him. There's coffee in a pot by the nurses' station in the back. We usually don't offer coffee, but since you're doctors, feel free to get some."

Dr. Warren looked at his clipboard.

"I've got another two kids to see down here and then I'm going to get a few hours of sleep—if I'm lucky."

Dr. Warren left with a conspiratorial smile, which I interpreted to mean that Pam and I both understood how precious sleep was for an ER resident and how lucky he would be to get even two or three uninterrupted hours.

Pam and I looked down on our child. An intravenous line was tapped at his wrist, a salt-and-sugar solution slowly dripping in. I counted the drops against the movement of the sweeping second hand of the clock and then stopped. I knew what infusion rate should be set for an adult who had not been eating or drinking for more than a day and who was "third spacing," in medical jargon meaning accumulating excess fluid in the third space of his abdomen. Had Dr. Warren set the rate correctly for a child of Steve's age and weight and condition?

I thought back to my nights as an intern and resident at Mass General. Sleep-deprived, inexperienced, constantly called to attend to several sick patients at once, I was terrified I would make a mistake. There were so many pressured moments when I worried I would miss a clue in a patient's history, overlook an important physical finding, misinterpret a laboratory test. Like the other trainees, I tried desperately to hide my fear by assuming the calm affect of an experienced physician in front of the patients and their families. We were supposed to be supervised by the chief resident, but the oversight was intentionally light-handed and uneven. You were a baby chick who had been pushed from the nest and had to learn to fly on your own. It was the way doctors had been trained for generations. And there was a macho pride in refraining from asking for backup.

It was a coup for an intern or junior resident, bleary-eyed and sleepless, to greet the chief resident in the morning having "held the fort" without asking for assistance.

In crises, I learned that my saviors were the nurses. They had seen generations of interns and residents come and go and were conversant with every detail of the patient's condition. They also were acutely aware of the interns' and residents' inexperience and the risk of error. Many were from working-class backgrounds, the first in their large families to complete a degree, and under different social and economic circumstances would have been likely to become doctors. Although you were the physician and they the nurses, you quickly learned to read the lightly veiled instructions within their queries and suggestions: "Doctor, shouldn't that blood transfusion be given more slowly for a thin elderly woman?" "We usually premedicate with antihistamines and steroids at higher doses than the ones you ordered before giving that antibiotic." "Hold the neck back a little more as you push down the tube into the trachea or else it may go into the esophagus." But there was no nurse to turn to, except McArdle, and she was in triage, no longer responsible for Steve's case.

"He looks so sick," Pam said softly. He did. Steve's skin was mottled, his eyes glazed and hardly moving. Only occasionally did he draw his stubby legs toward his chest, but weakly.

"And he seems to have more fever. Here, feel his head."

Pam's soft hand guided mine to Steve's crown. He was hotter than before, although he had been given Tylenol before the barium enema.

A sense of dread grew within me. Steve was deteriorating. The conclusion about Dr. Warren that I had read in Pam's eyes echoed loudly in my mind: *He's not listening carefully; he wasn't paying close attention.* Now Dr. Warren had asserted that, "in my experience," Steve could safely be observed overnight. And after two other children were seen, he hoped to get some sleep.

"I don't trust him," Pam stated firmly. "It doesn't feel safe to wait."

I nodded silently in agreement. But what could we do? If we insisted on going to the operating room, it would be Dr. Warren wielding the scalpel along with his nameless intern who had been looking for a good case.

I felt off balance and reached out to clutch Pam's hand, using her as ballast to gain equilibrium. For a long, terrifying moment my mind was blank. I then forced myself to think. Pam had trained at Harvard Medical School. She had rotated through the Children's Hospital. But she was only a student then, and that was many years ago. She had no personal contacts. I had worked only in a basic science lab, as a researcher, not in the hospital . . .

"David Nathan?" I blurted before my thought was finished.

Pam looked uncertainly at me.

I had bumped into David bicycling with his wife, Jean, along Memorial Drive a few days before we left for California. He had wished me luck in my new job at UCLA, and I had thanked him for the basic science training I got in his department. I knew he lived in Cambridge but also recalled that he was an avid sailor. He spent most summer weekends on his boat moored in Nantucket Harbor.

It was a long shot but the only one we had. I had three quarters in my pocket, the change from the Allston–Cambridge toll on the Mass Pike.

"In Cambridge. N-A-T-H-A-N, David and Jean. Don't know the address," I told the operator from the pay phone in the ER waiting room. I cupped my hand over my exposed ear, muffling the sounds of the crying children and frazzled parents camped out around me. Most doctors, for reasons of privacy, don't list their home numbers, only their office numbers.

"He's listed."

Thank God, I said to myself.

I wrote the number on a scrap of paper towel I had taken from the ER room. *Please be home.*

I looked at my watch. A quarter to one. I let the phone ring for what seemed like an eternity.

"Hello? Uh, hello?"

It was David's gravelly baritone. The tentative timbre indicated I had awakened him.

"David, it's Jerry Groopman. I'm sorry, I know it's early, one A.M., but I'm in the Children's ER. I need your help."

He paused briefly before replying.

"I thought you were in California . . ."

"We're back."

I quickly explained the events of the past day, giving only the essential clinical details, the confirmed diagnosis of intussusception, fever, elevated white count, Steven's flagging strength. I then stated our lack of confidence in the resident, and why.

"I'm calling Ray Levy," David tersely replied. I knew of Dr. Levy only by reputation. He was spoken of in terms of awe, one of the most renowned pediatric surgeons in the country. "I know he's in town. He lives in Weston. Ray'll contact this Dr. Warren in the ER."

I thanked him, multiple times, my eyes filling with tears.

Not more than ten minutes later, Dr. Warren entered the exam room. I stood up from my chair to meet him, holding his gaze firmly in mine.

"Dr. Ray Levy is driving in from home," he flatly informed us.

His expression was a mix of resentment and fear, but I no longer cared what Dr. Warren felt. We had done what we had to do. And the fact that Dr. Levy was coming in meant he did not necessarily agree with Dr. Warren's plan to get some sleep while Steve waited until the morning.

Ray Levy stood with us in the exam room. A tall, portly man with fine wisps of gray hair and delicate hands, he greeted us without a trace of annoyance at being called in the middle of the night and on a holiday weekend. He took the history again from Pam, examined Steven himself, and then intently studied the X rays on the light box outside the ER room.

"He needs to be explored," he soberly told us as Pam's hand met mine in a tight grasp.

Dr. Levy said that although we were both doctors, he wanted to clearly explain the need for emergency surgery. The rising fever and elevated white cell count indicated that the bowel might be starved of its circulation and could soon rupture. If ruptured, it would spill its contents into the abdomen, causing peritonitis, and potentially fatal shock. Moreover, the appearance on the X ray of such large ballooned loops of intestine and the failure of a forceful enema to open the blockage suggested that the intussusception would not resolve spontaneously. Waiting until the morning only increased the risk of rupture.

We both signed the consent form for the surgery, not really reading it. It was enough to understand the risks if the operation were not done: that Steve's bowel would burst, and he might then die.

Steve was taken from us again, Dr. Levy walking with the nurse, guiding the stretcher himself to the OR. We took the elevator up to the waiting area.

I sat next to Pam in silence and closed my eyes. A series of flickering images appeared, like those from silent films. Steve dead, his lifeless form in a closed, rectangular pine coffin. But I could see inside. He was clothed in a white shroud, according to traditional Jewish custom. I then watched the coffin lowered into the moist ground in an overcrowded cemetery in Queens, next to my father. Dirt was shoveled over the coffin. Then the drab Waltham home, mirrors covered, Pam and I seated on low stools, clothes rent—shiva—the mourning of our firstborn son.

I forced the pictures to be erased from my mind. They were replaced by cold thoughts. We might have other children but never Steve. His absence would fill every event in our lives. His name would never connote wisdom and courage and compassion. It would be a lacerating echo in our hollowed hearts. I knew I would think of him when I lay on my deathbed. Pam and I would become like other parents we had known who had lost a child: that child was a vital part of you, and when he died, that part of you died.

Please, God, save the life of my beloved son.

The air was still and moist as the first crimson rays of a rising July sun emerged over the horizon. Its glow spread through the OR waiting room and cast an aura on Dr. Levy as he stood before us.

"It was fortunate we operated," Dr. Levy said as I drew Pam close to me. "The bowel was ashen, moments from perforating. Fortunately, we were able to undo the intussusception without a resection. His intestine is completely intact. There was no tumor or other apparent cause. He'll be in the ICU on antibiotics for a few days but should be home by the end of the week."

I began to cry, as did Pam. There were no words to convey how I felt for Dr. Levy at that moment. He was the angel who descended from heaven and breathed life back into our stricken son. I managed a repetition of "Thank you." He clasped my hand and seemed almost embarrassed by the gratitude.

"I'll see him midday. I have another case scheduled for shortly after six."

I have told this story many times to many people. Sometimes it's during the summer, around the July 4th weekend, when the memories flood back. We are swimming as a family at a pool in Brookline and the long horizontal scar that indents the flesh of Steve's abdomen triggers the association. At other times, it is in synagogue, when I am with my children, and we reach the point in the service where prayers are offered for the healing of the sick. But most often it is on teaching rounds, with young doctors in training. We talk about mistakes, acknowledging that we are all fallible, and that a physician must, at least once, and probably many more times in his career, be in error.

I tell my students about that night with Dr. Warren in the ER, and about Dr. Burgess. I then recount how the surgical interns and residents who worked beside me during my training at the Massachusetts General Hospital had a closed-door session where they confessed their errors—or at least the ones that they recognized. Their mistakes were classified as ET or EJ, for "error in technique" or "error in judgment." An ET was of less gravity than an EJ. An ET was addressed with the maxim that "practice would make perfect." An EJ reached into a more profound dimension of doctoring. The senior attending staff would seek to identify the roots of the resident's error in judgment. It was rare that lack of specific knowledge caused the EJ. Rather, it occurred as it did with Dr. Warren. Whether because of fatigue, distraction, or arrogance, he did not listen carefully to Pam. Not during the history, and not later, after the diagnosis was made. He did not respect her instincts about her child. He then made a critical decision, one that nearly cost our son his life, by pretending he had the experience to do so, without seeking other opinions. This type of EJ was not the sole province of Dr. Warren. Dr. Burgess dismissed Pam's observation of melena and closed his mind quickly to the broader range of possible diagnoses.

Lay friends hearing this story uniformly state: "What if you and Pam weren't doctors?" There is a hard inescapable truth to that. We knew that July is a problematic month, and that there are interns and residents who are good and bad, sharp and sloppy. I had a professional connection with David Nathan that brought in Dr. Levy. But we didn't think to use it until we stopped being passive and challenged Dr. Warren's judgment. What we share with all patients and their families is the sense of when something is seriously wrong with someone we know so intimately. Technical knowledge, that we were more expert than Dr. Warren and could second-guess him, was not the pivotal factor. We doubted his competence because we saw that he was not focused on Steve, that he was tired, distracted, and eager to sleep, and that his decision was couched in the arrogance of limited experience.

These points were brought home to me recently in speaking with Ellen O'Connor. Ellen is in her late twenties, a soft-spoken woman with gray-blue eyes and a ready smile. She is a data processor on the staff of a firm whose services I use. Her husband, Michael, is a salesman. In the autumn of 1997, the O'Connors' first child, Sandra, at the age of six months, developed fever and diarrhea. The local pediatrician was concerned about the possibility of intussusception and sent them to a hospital emergency room. There, the intern examined Sandra and obtained an X ray of her abdomen.

"He told me that he 'thought' the X rays suggested an intestinal blockage and Sandra needed exploratory surgery," Ellen said, her face tightening with the memory. "I asked him whether he *thought* or he *knew*. He couldn't give me a straight answer."

The O'Connors waited several hours for blood tests to return and for the surgical team to be assembled. During that time, Sandra improved. She felt less feverish to Ellen, and her disposition turned from cranky to calm.

"I told the doctor that she was herself again, but he didn't listen to me. He insisted that Sandra be taken for exploratory surgery."

Ellen and Michael stood their ground. They are self-made people from working-class families and not easily cowed.

"I'm not a doctor or a nurse," Ellen continued. "But I'm a mother, and I know my child, and I felt deep inside that she was okay."

After much argument, the intern told the O'Connors that if they refused to have their baby taken to the OR, they would have to sign a legal release assuming responsibility for this decision.

"I asked what other tests could prove his suspicion. He told me an ultrasound. But it was expensive and required a specialist to come in and do it. It was late evening by then, but I said that no way was I going to back down. The specialist was called. He gave us a really hard time," she said, shaking her head.

To the doctors' surprise, the ultrasound showed the intestines were moving normally. There was no intussusception. Sandra had a routine viral gastroenteritis.

"It was a hard decision for people like us to question the doctors," Ellen said. "But we needed to be sure they were right. It was my baby."

A Clinical Enigma

"I just don't know what to do, old boy," James Hunt said. An Englishman some fifteen years my senior, he was a general internist in rural New Hampshire. "Our friend Peter Emery is a clinical enigma. He has some kind of lung process and a bizarre fever pattern. One day he's at 104 and looks like bloody hell—ashen, blood pressure low, breathing hard, needs oxygen, and I think I'm going to lose him—and then the next day—fine, perfectly fine, no fever, bright and cheery, full of energy, like he's ready to take my horse to the polo match."

It was the Thursday before Labor Day weekend 1998. Some nine months earlier, I had assumed the role of consulting hematologist on Peter Emery's case. Dr. Stephen Robinson, an eminent blood specialist at my hospital, was the original consultant, but Steve had developed cancer. A dedicated and determined doctor, he continued as long as he could but finally acquiesced. He distributed his practice to his colleagues based on the patient's diagnosis and predicted compatibility. I was asked to care for Mr. Emery.

Before I met Peter Emery I met his medical records, two thick volumes.

In 1993, then fifty-six years old, he had complained to James Hunt that his feet and hands were swelling. James performed a detailed evaluation and found that Mr. Emery's kidneys were malfunctioning. James Hunt astutely determined the underlying problem came from outside the kidneys. Mr. Emery had a blood cancer called myeloma. Here, a type of white cell, called a plasma cell, becomes malignant and produces fragments of antibodies in great excess. These antibody fragments are unnecessary, made in an uncontrolled fashion, like a printer stuck in overdrive spitting out sheet after sheet of the same page. Peter Emery's kidneys were clogged with the antibody fragments and his bones riddled with myeloma. He was referred to specialists at our hospital.

Dr. Robinson administered Alkeran, an oral chemotherapy, and prednisone, a corticosteroid, for the myeloma. Dr. Franklin Epstein, a kidney specialist, advised on the water, protein, and electrolyte balance. After a year of intensive treatment, the blood cancer remitted and the kidney function returned to normal. Mr. Emery, semiretired on his family farm in New Hampshire, returned to chopping wood, digging wells, improving his golf swing, and taking long walks with his wife, June.

Was the current clinical enigma a sign the myeloma had returned? Or was it related to Peter Emery's second disease, one that was "iatrogenic," meaning brought on by his physicians' treatments?

The Alkeran chemotherapy that eradicated the myeloma had injured Peter Emery's bone marrow. This resulted in widespread scarring, or "myelofibrosis." The marrow cells choked under the blanket of scar, so little blood was produced. Peter's immune defenses were low; this could permit long-dormant infections to reawaken.

Peter had been the chief executive of an international oil company in Italy. He traveled widely, not only in Europe but also in Africa, South America, and Asia. I asked James Hunt about reactivation of exotic microbes as a cause of Peter's "bizarre" fever.

"Peter had malaria in Senegal and some type of dysentery in Venezuela," James said. "We have cultures cooking. The thick smear for malaria was negative.

"Peter wants me to remind you of our deal," James added.

Peter, half jokingly, had said that each year his doctors kept him alive, they and their spouses were invited to accompany June and him to Italy. Peter would treat everyone to one of the "most spectacular dinners imaginable."

I was doubtful from the start if I would ever attend such a dinner. The treatment options for myelofibrosis are painfully limited, and the scarring is generally progressive. The patient, unable to form blood, dies of infection from lack of white cells and hemorrhage from lack of platelets. A raging fever of unknown cause solidified my sense of doubt.

I first met Peter Emery on a snowy January in 1998. He was nearly my height, six foot five, a handsome man with rugged features and deep-set blue eyes. When we shook hands, his grip was strong but uneven, and I recalled from his chart he had lost two fingers on active duty after graduating from the U.S. Naval Academy in Annapolis. June stood at his side, a diminutive woman with salt-and-pepper hair and soft hazel eyes.

"Dr. Robinson affirmed you're the man with the most experience with this scar condition," he said. "I assume you've reviewed my chart and James Hunt has updated you on my latest blood tests. So let's get down to business. What exactly is going on here?"

He smiled, a broad, warm smile that softened his no-nonsense gaze.

I explained that the injured marrow cells released inflammatory substances. These inflammatory substances fostered the growth of the scar. The more scar, the more the marrow cells struggled and became inflamed. It was a vicious cycle triggered by the chemotherapy five years before.

"No reason to look back—we had no choice about what had to be done. I'm the kind of person who tries to fix things. Even if there's only a small chance of success. And don't worry too much about side effects. I can put up with a lot. I worked in some pretty tough terrain."

We first tried interferon, which had some success at the Mayo Clinic in slowing the scar. But after three months of regular injections, we were forced to conclude it failed. Peter was still severely anemic and dependent on blood transfusions.

After interferon failed, we met again. Peter asserted he was not ready to give up.

"Our daughter, Faith, searched the Internet under 'myelofibrosis,'" Peter said. "She forwarded this printout to me."

There were some reports from Britain and Italy of a few patients benefiting from corticosteroids, like prednisone. Most did not improve. It was presumed these rare successes occurred because corticosteroids are potent anti-inflammatory drugs and reduced the inflammation in the marrow.

The rest of the material was about bone marrow transplantation. A few cases of myelofibrosis reversed after it.

Faith told Peter that the procedure would work by first wiping out all his blood cells with lethal doses of radiation and chemotherapy and then grafting new blood cells from a compatible donor. These healthy donor cells had not suffered the injury from Alkeran and thus did not release the inflammatory factors that cause scarring. The marrow would heal.

"But it's not so simple," I explained to Peter. First, in many patients in whom it was attempted, the scarred marrow was an inhospitable home for the infused donor cells; they could not settle into supportive niches, due to the extensive distortion of the cavity. The graft did not "take," in medical jargon, and the patient died.

"That's one risk. Any others to consider?"

There were. Marrow transplantation was more difficult in older patients. The successes cited in the Internet material were among people all younger than forty. In one's sixties, the rate of complications was higher. The procedure had profound toxicities on heart, lungs, bowel, liver, and kidneys.

"My kidneys were seriously damaged before, from the myeloma," Peter interjected. "Dr. Epstein said they're still fragile."

I acknowledged that made them more susceptible to subsequent damage. Moreover, there was so-called graft-versus-host disease. If the donor cells do "take" in the emptied marrow cavity, they grow into all the components of the blood and immune system. The reborn immune system from the donor perceives its new host as foreign and attacks the host's liver, bowel, and skin. Powerful immune suppressive medications are needed to temper this reaction of the graft against the host; in some cases, there is chronic disability from the damage to these organs, with prolonged hepatitis, diarrhea, and skin rashes.

"I appreciate your candor and the clarity of your explanation," Peter Emery said. He paused and then locked his sharp eyes on mine. "It's important that you know quality of life is foremost for me. One of my sons, Jared, was a nurse in an ICU for many years."

I hadn't been told that by James Hunt.

"Jared went to Dartmouth and then left when he wanted to be a nurse. He's a kid who marches to a different drummer. That's the way life is, I guess. You learn to accommodate different kinds. Jared told us how people are kept alive beyond what's reasonable or natural. That's not for me."

The message was clear.

"I'm going to pass on the marrow transplant," Mr. Emery said as he folded the computer printout neatly in half. "As to other options you mentioned, I've had prednisone before, of course, for the myeloma. Not much fun. Those corticosteroids make me edgy, unable to sleep, and constantly hungry. I'm not one to kick the dog, so I'll plan on cutting a few more cords of firewood for the winter to get the aggression out. And June will have to lock the larder when I'm in the mood for a raid."

I prescribed prednisone to be taken every other day, at moderate doses. He had the expected side effects and stoically coped with them. I knew it was a very long shot, and the faxes from James Hunt confirmed that: no improvement in blood counts; regular transfusions still needed.

"I'll hold on to him a bit longer, until the cultures ripen," James concluded that Thursday before Labor Day. "All we have are these fuzzy changes on the chest X ray. There's no sputum to speak of. Doesn't look like a typical community-acquired pneumonia. I've got him broadly covered with empiric antibiotics, but I soon may need a little help here."

I said that he should not hesitate to transfer Peter to Boston. We exchanged private numbers, since I was not on call; James could contact me at home, and I would facilitate the logistics.

The call came that Saturday night. James Hunt had spent the better part of the day in the ICU trying to maintain Peter Emery's vital signs. His fever reached beyond 104, his arterial oxygen plummeted, and his blood pressure was unstable.

James said he needed the night to stabilize Peter and ensure it was safe to make the four-hour trip by ambulance. We agreed to talk first thing in the morning; meanwhile, I would alert the emergency room to expect the transfer.

I left my house that Sunday morning an hour before the ambulance was to arrive from New Hampshire. I wanted to go over the case again with the hematology fellow and the resident in the ER. There would be two aspects of Peter Emery's management: the first, to support his heart, lung, and other vital organ function; the second, to identify the root cause of his fevers and "lung process."

Carolyn Krasner was the fellow, and Mark McDaniel the resident. Each had already reviewed the chart and informed the pulmonary team that Mr. Emery might need to be intubated and placed on a respirator as soon as he "hit the ER." They had also secured a bed in the ICU and given the medical team up there a synopsis of the case.

The ER was quiet that morning, a few patients still under observation from the night before. Mark McDaniel said most would be going home with the usual: casts for broken limbs, a score of stitches for a gash, antacids when the pain proved to be indigestion and not a myocardial infarction.

Our conversation was cut off by the sharp wail of the sirens. My pulse quickened and my muscles tensed. The three of us stood in unison, eyes trained on the electric doors of the ER portal. The two nurses at the triage desk stood as well; one would receive the transfer forms from New Hampshire, the other would assist in assessing and stabilizing Peter.

The automatic doors rapidly separated and the blue-uniformed ambulance drivers wheeled in the stretcher. I was on the balls of my feet, poised to run to meet them.

I lowered my heels to the floor.

Peter was holding the first section of the Sunday *New York Times*, glasses on, hair neatly groomed, in blue cotton pajamas and a white terry-cloth robe. His signature smile creased the corners of his face.

I exhaled and felt my heart slow and the tension in my arms and legs gradually diminish. Mark McDaniel looked at me wide-eyed.

"Maybe Mr. Emery can share the *Times Book Review* with me," Carolyn nervously chuckled.

We reviewed with Peter the events of the past hours.

"The siege lifted around two A.M. Don't even need this oxygen," he said, lifting the prongs away from his nostrils. "The ambulance men insisted I keep it on during the ride."

He explained that June was driving down herself and would arrive in an hour or so.

His physical examination revealed harsh breath sounds at the base of his right lung. The chest X ray from New Hampshire showed, as

James Hunt said, a ground-glass appearance instead of the normal blackness signifying air in that area. There were no other abnormal findings.

Mark McDaniel, the resident, drew new blood cultures and standard chemistries. Carolyn canceled the ICU bed and arranged for admission to a regular medical floor.

"Bloody hell, I must have lost my nerve," James Hunt exclaimed when I called him. "Terribly sorry to send him in and ruin your Sunday."

I said it was always hard to know which way a patient was headed. James had covered a bacterial pneumonia with empiric antibiotics. Perhaps Mr. Emery's course was simply one of stuttering improvement. It was also possible the airways were blocked by plugs of mucus, so the microbes were released periodically into the bloodstream as the mucus plugs shifted. This could account for the episodes of high fever and falling blood pressure. Over time, with James's cocktail of antibiotics, the lungs would fully clear.

I imagined myself as James and felt dwarfed by his tasks. He was alone in a remote part of New Hampshire, with no cadre of specialists, as I had, to verify his clinical impressions, and no fellows, residents, and interns on call to execute the demanding labors of care: the spinal taps, the nasogastric tubes, the intravenous catheters. James was trained under the British system, where bedside skills were paramount. Listening attentively to a patient's history, followed by a meticulous physical examination, formed the core of classic diagnosis and treatment. Sophisticated blood tests and X rays were used only when required, not as a knee-jerk response to every complaint.

The transfer also said something admirable about James: that when he felt he had reached his limits, he did not eschew admitting it and readily sought help.

I left the hospital, went for a long and much-needed swim, and returned home. As I settled into the Sunday *Times Magazine*, I was paged.

"Peter Emery is on his way to the ICU," Carolyn Krasner told me.

In the last hour, his fever skyrocketed, his blood pressure fell, and his breathing became labored.

I arrived at the hospital in less than fifteen minutes, the traffic being sparse. Peter was positioned at a 45-degree gatch, a large green oxygen mask covering his nose and mouth. His exposed arms were marked by a reticulated pattern of bluish gray, a sign of imminent circulatory collapse.

The residents were infusing large volumes of saline to support his blood pressure. The pulmonary team stood at the bedside, debating whether to intubate him now or wait to see if the crisis passed.

I grasped Peter's limp hand. He turned his head toward me. His nostrils were flared and his mouth open as he gulped air. His usual calm eyes were wide with alarm.

"We're getting on top of it, Peter. We'll lower your temperature and support your pressure. Just hold on."

I wasn't sure of this, but he nodded appreciatively.

The nurses bathed him in alcohol to bring down the fever. An intern held his forearm, cleaned the area with iodine, and punctured the sterilized vein for yet more blood. A nurse swabbed the end of his penis, a urinary catheter held in her other hand; an accurate assessment of outflow was needed to be sure his kidneys weren't shutting down.

I explained to Peter these multiple interventions and then said I would step out for a moment to update James Hunt.

"If you and he pull me through . . . my treat to two dinners."

James was at home.

"Well, I guess I was right to make the transfer. But what the hell is going on?"

We began to consider unusual diseases. Peter's immune system was not normal, the scarred marrow releasing white cells that weren't fully competent to protect against fungi. Perhaps one of these organisms had taken root deep in his right lung. But that didn't explain the episodic nature of the illness. That pattern might be caused by walled-off areas of infection, meaning abscesses, that released such microbes in spurts.

"Keep me posted, old boy," James said.

I stayed in the hospital, contacting specialists and awaiting the outcome of the current episode. It passed, and some three hours later, Peter was seated in bed, watching the eleven o'clock news.

"Terrible, this whole Lewinsky business," he said. His face was wan, and I noticed how much effort it took for him to bring the glass of ginger ale to his lips. "The man lied, of course, but this is not 'high crimes and misdemeanors.' It's stupidity. The country has better things to do than cascade down this slippery slope of impeachment. Just censure him and get on with business."

We watched a little more of the program, and then Peter shut it off. He turned to me, his rugged granite features, bold chin, and bright eyes all sagging under the burden of what he had endured.

"Don't hold back on me," he said. "Probe and prod and do whatever must be done to get to the bottom of this."

Tomorrow was a bronchoscopy. The pulmonary doctors would place a bronchoscope, a fiber-optic instrument that worked like a telescope except you could see around corners, into his lungs. It would be uncomfortable, but it would allow us to visualize his airways. I explained my mucus plug theory, that perhaps concretions of mucus were acting like ball valves, blocking the clearing of sputum, perpetuating a fungal pneumonia.

"Could the ball valves not be plugs of mucus but lung cancer?"

I said it could be, but we thought that unlikely. He hadn't smoked.

"But I've been around the Ukraine, including Chernobyl, and God knows what radioactivity was cooking in the navy in the late fifties."

The bronchoscopy would detect such a cancer. As for my other theory, of an abscess, a CAT scan of his thorax would show that. A bone marrow biopsy would also be done. Mycobacteria, organisms like tuberculosis, as well as fungi, could be identified in the marrow, sometimes more easily than in cultures of blood or sputum.

"And if you come up empty-handed?"

Then we would take the next step: an open lung biopsy. It was surgery, necessitating general anesthesia, his chest opened and a piece of tissue obtained for pathological examination and microbial culture.

"Whatever it takes, Jerry. Whatever it takes."

The bronchoscopy showed his airways to be open, without any obstruction by mucus plugs or cancer. The CAT scan of the thorax did not detect an abscess or other cloistered pockets of infection. The marrow showed extensive scarring but no microbes.

His fevers ranged between 101 and 103, and his respiration required the oxygen mask. But there were no further drops in blood pressure.

"So where are we today, old man?" James Hunt asked a week after the transfer.

I wasn't sure. Peter Emery's illness still was without an explanation. Consultants from nephrology, pulmonary medicine, and cardiology had seen him. I had presented his case to two specialty clinical conferences: one in hematology, the other in infectious diseases. These conferences occur each week, bringing together senior specialists from the affiliated Harvard hospitals to analyze the "toughest cases." The principles of differential diagnosis are applied, meaning the creation of lists of possible causes, beginning with the common and ending with the arcane. Although no one could say what he had, the suspicion centered on a fungus. Peter was immune compromised, because of the underlying marrow disease and the treatment with corticosteroids. Fungi love this mix of immune deficiency and steroids, and are notoriously difficult to grow out in culture. Viruses were a second possibility, also vexing to identify, since there are so many of them and the techniques to culture them inefficient. But the height of the temperatures, the episodic nature of the crises, and the three-week-long duration of the illness all made viral pneumonia less likely.

We looked again for reactivation of tropical diseases, like malaria and rickettsiae, and that search proved negative. We investigated his endocrine system, seeking a hormonal cause for the instability in blood pressure. Peter was on prednisone for several months; this suppresses the adrenal glands. If the adrenals cannot respond to the stress of infection or inflammation by releasing their own natural mix of steroids, then blood pressure cannot be maintained. But the signature changes of low sodium and high potassium, marking such adrenal suppression, were not present. Furthermore, James Hunt had thought to give Peter replacement adrenal steroids in New Hampshire, and this prophylaxis had not averted the crises.

"Could it be some arcane inflammatory process, due to the aberrant white cells of myelofibrosis?" James asked when I reported on our failure to decipher the illness.

James's query was, of course, a diagnosis of exclusion, meaning we would rule out all secondary causes and attribute the problem to an inherent dimension of the underlying marrow disease. There were, I told him, case reports of inflammation occurring in the lung. But I was wary of these anecdotes. Patients with myelofibrosis were regularly transfused and thus susceptible to infection with blood-borne viruses, like hepatitis. Hepatitis virus could trigger inflammation of small blood vessels in the lung, as well as skin, kidneys, heart, and brain.

"But we checked for all the hepatitis viruses, now didn't we?" James countered.

Indeed, we had. As well as other systemic viruses that can travel in blood products, like HIV, HTLV, and cytomegalovirus. The pulmonary consultant, I said to James, concluded it was idiopathic.

"I detest that word 'idiopathic,'" he spat. "Sounds like it comes from the words 'idiot' and 'pathetic.'"

I laughed nervously and affirmed I felt the same way.

Idiopathic means without known cause. It is a "wastebasket" term applied to disease conditions whose origins were not known. But I knew there was more danger than humor in invoking it. By designating a syndrome as idiopathic you indicate you are satisfied with your ignorance and ready to stop searching further for a discrete cause.

"You have no choice, old boy," James concluded.

We had to proceed to an open lung biopsy.

I went to Peter's room in the early evening. There was a panoramic view of downtown Boston, the last rays of the setting sun igniting a blaze of crimson over the mirrored Hancock Tower in Copley Square. I pulled the blinds down to mute the brilliance of the light.

Peter was soaked in his pajamas, his fevers just shy of 104 all afternoon. He was breathing hard, despite the high-flow oxygen mask.

"You know . . . when I was in management," he said in forced phrases, "and we faced a knotty problem like this . . . one where there was no obvious solution . . . I'd bring together the best minds into one room . . . and lock the door . . . until we came up with a satisfactory answer . . . or a new strategy."

I explained I had done this by presenting his case to two such Harvard convocations in infectious diseases and in hematology.

"How about calling in . . . someone from outside . . . maybe the Centers for Disease Control?"

The Centers for Disease Control is the federal body based in Atlanta responsible for epidemics and the monitoring of the public health. Peter's question was a polite way of saying that perhaps a specialist outside the Harvard fold would have more insight into his condition.

I replied that our chief of Infectious Diseases, Dr. Karchmer, who was one of the leading consultants in the nation, had spent several years in Atlanta working there. He was perplexed by the case as well. And I had also contacted colleagues at other institutions: Dr. Stephen Nimer, who headed the department of blood diseases at Memorial Sloan Kettering and was expert in marrow failure states, and other seasoned hematologists at the NIH, Mayo Clinic, and the Fred Hutchinson Cancer Research Center in Seattle.

"June . . . make a note . . . to reimburse Jerry for all those phone calls."

June smiled and then asked matter-of-factly, "Tell us more about the open lung biopsy."

Her worry was evident in the flatness of her tone.

I knew they had been married some thirty-six years. She had borne four children and followed Peter around the globe. In each place, civilized or primitive, she set up a tight household, arranged for the children's education, joined the local church, cultivated a circle of new friends. She was the tactician, he the strategist.

I outlined the risks of general anesthesia, and then the issue of "pneumothorax," or air in the chest. After the piece of lung tissue was taken, a hole in the lung was left. Air built up between the exterior of the lung and the inner chest wall and collapsed the lung. A chest tube would have to be inserted through the ribs to suck out the air. This was painful—it was like being stabbed in the chest with the knife left in. The stabbing chest tube might need to be kept in place for a few days, until the biopsy site healed and the lung was sealed. And, as with all procedures, but particularly in light of his underlying marrow disorder, bleeding or infection needed to be considered.

"But this biopsy . . . is the critical final test, isn't it?" she said.

I replied that it was.

Three days later I sat at his bedside with arguably the worst possible news—the biopsy showed extensive inflammation, and some necrosis, or death, of lung tissue, but no microbes or other clear reason to explain the condition.

Peter was in a fetal position, knees flexed on his side, the pain of the chest tube too great to permit him to lie flat on his back. He accepted only the minimal amount of narcotics for the pain, determined to stay alert and engage in conversation and watch the TV news. But, he admitted, he was too uncomfortable to read.

He was frustrated, he said, stuck midway through a tome on British naval strategy in the Second World War.

"I studied engineering at Annapolis. . . . I thought that was a sure ticket to a career. . . . But my love has always been strategic thinking. . . . There was too much emphasis on technology in engineering. . . . Technology is always secondary to strategy."

I sat silently. I now had neither a clear strategy nor novel technologies to apply. With the biopsy results in hand, I called back my three colleagues, one at Sloan-Kettering, one at NIH, and one in Seattle. We all agreed that while the inflammation and necrosis might be due to his underlying myelofibrosis, it still was an unsatisfactory assumption. Fungi regularly caused tissue destruction and might have been missed by the surgeon's blade. We were still awaiting electron microscopy to identify organisms not seen by standard light microscopy. This would take a few more days, but I sensed we would not find them. We were slipping into the dark abyss of idiopathic.

"We'll continue the supportive measures," I said to Peter and June, "oxygen and prednisone. Amphotericin, the treatment for a fungus, is very toxic. I want to wait before empirically starting it. Amphotericin can cause kidney damage."

"Sometimes it's hard to locate the enemy," Peter gasped, "even with the best radar—even after dropping behind their lines." He tried to assume an air of confidence. "Keep on looking. . . . You'll find it. . . . It just may take time."

⁓

"Why wasn't I called?" I bristled.

"It was two A.M. when he went back to the ICU," the resident explained. "The ICU attending came in. We figured we'd wait to call you until morning."

I looked at the bedside digital clock. It was 7:20 A.M., Sunday, September 27. I was reassured to hear that the director of the ICU personally came in to manage the downturn but still was annoyed that I was spared the call. I realized my presence was not essential. Everything had been done expeditiously and efficiently. My response was an expression of my growing frustration about my failure to diagnose his case.

"The downturn didn't occur with the amphotericin, did it?"

The intern said no. Amphotericin, the toxic antibiotic for deep-seated fungus, was begun some two days before, in desperation. As the week progressed, Peter worsened. The chest X ray showed that his left lung was affected by whatever was damaging the right. This had forced my hand to begin amphotericin despite his fragile kidneys.

I entered the ICU through the side door. June greeted me at the portal to Peter's room. Faith was seated at his bedside. She had flown up from Washington on the first shuttle. I had met her before; she accompanied her father to the clinic once when June was unavailable. Jared, the son Peter said marched to a different drummer, was standing in the far corner. A tall, lanky man in his early thirties, with curly dark hair and an olive complexion, he contrasted with the creamy complexion, blue eyes, and flaxen hair of his sister.

Peter's lungs were unable to capture enough oxygen and release enough carbon dioxide despite the mask. So a tube had been inserted into his trachea, and he was connected to a mechanical respirator.

A large computer was positioned next to Peter's bedside. It displayed in real time his vital signs—respiration, pulse, temperature—and in adjacent columns, the most recent laboratory results—blood oxygen and carbon dioxide; sodium, potassium, calcium, and other electrolytes; hemoglobin; kidney and liver function tests; and internal pressures in his heart from a catheter placed last night in his right ventricle.

I hugged June. She looked darkly at me. Faith met my gaze at eye level. Her expression was uncertain. Jared had an intense, serious mien, studying the computer screen along with me.

Peter was asleep, exhausted from the travails of the night.

"Should we step out?" Jared asked.

We retired to a conference room adjoining the ICU. Carolyn, the nurse who was attending Peter during the day shift, joined us. The room was used for rounds and had a light box on the wall. Peter's chest X ray was mounted on it. The changes in his right lung had become more pronounced, and the ground-glass haze on the left extended halfway up from the diaphragm. There was little lung on either side unaffected.

I sat at the head of the metal table, with Carolyn and June to my left, Faith and Jared to my right.

"Matthew is on his way down from New Hampshire, and Peter Jr. is flying in from New York," June said of her two other children. "I'd hoped they'd be here already."

She glanced at the large clock on the wall behind me. I looked reflexively at my watch: 10:16.

"You all understand the events of last night?" I asked. Before sounding out their wishes I needed to be certain everyone was at the same level of knowledge.

"I think we do," Jared answered. June and Faith silently nodded. Jared continued. "And we have real concerns about what is happening."

His face was grim.

"What's the point here? Dad has end-stage myelofibrosis. You've got him on a ventilator. He's at a maximum FiO2. His circulation is dependent on a pressor, Levophed. And his creatinine is climbing. Next you'll have to begin dialysis."

I glanced at June, then at Faith; each wore an uncertain expression. Carolyn looked perplexed.

"Jared was an ICU nurse," I explained to Carolyn.

I translated for June and Faith: maximum FiO2 meant the highest level of oxygen we could safely deliver by the respirator; Levophed was a "pressor," meaning an Adrenalin-like drug that supported blood pressure when the body's circulatory mechanisms failed; "creatinine" referred to a blood test that rose with kidney failure and would guide the decision about dialysis.

"His kidneys were damaged by the myeloma. Are they failing again?" June asked. "He didn't want dialysis then, to live hooked up to a machine. Please don't do that."

The discussion was moving ahead too fast. I needed to rein it back in.

"Let's first focus on what is happening acutely, meaning last night's crisis. His chest X ray shows extension of the process from the right lung into the left. His inability to get enough oxygen likely caused the drop in blood pressure, and that reduced the circulation to his kidneys, reflected in the rising creatinine."

"It's always a domino effect," Jared interrupted. "And it may not be that particular scenario. Yes, the chest X ray confirms worsening in the lungs. But you started amphotericin—empirically, without documentation that he actually has a fungus—and that drug is toxic to the kidneys and causes drops in blood pressure." He paused and said with obvious irritation, "It's not worth becoming hung up in the details. The big picture is very bad. We've discussed it as a family. We want him taken off the respirator. *Now.*"

Carolyn looked at Jared with alarm.

"But he'll die within minutes," she said.

"We realize that," Jared shot back.

He marches to the beat of a different drummer.

I wondered whether there was a subtext beyond the apparent. My mind filled with possibilities. Did Jared want his father dead? Would that be the ultimate victory in a long struggle between two clashing personalities? Or could there be other reasons, issues of inheritance? Or was Jared expressing the anguish we all felt, and simply and deeply believed ending his father's suffering was the ultimate act of a son's love?

"I don't feel comfortable disconnecting him now," Carolyn stated emphatically.

"Nor do I," I echoed. "He's been on the amphotericin for just two days. If it is a fungus, we might not see a response yet. And, most important, nothing has happened that is irreversible. His lungs may improve, his blood pressure may return, and his kidneys may recover—without dialysis."

"But what kind of life will he have, Jerry?" June asked, her voice choking. "Tell me—truthfully—don't spare me the details. What are we looking at if this all does reverse? What will *he* have to look forward to over the next weeks and months? Peter had such a good summer. He worked on the well, played some golf, entertained friends, even chopped a few cords of wood. Will he return to that? Won't he be terribly debilitated, his lungs weak, his bone marrow more scarred? It will be more and more transfusions, tied to an oxygen mask. And then it will be the next infection and back in the hospital."

"It's hard to say," I replied. "He might rally. And if the amphotericin works, he might return to a better level of functioning."

"You really believe that?" Jared charged. "I don't. I worked in an ICU for nearly two years. He'll have residual lung damage. He won't go directly home. He needs nurses, respiratory therapists. Dad will be in some rehab facility for months. Maybe eventually he'll get out, but more likely his weakened immune system will mean another pneumonia, or whatever lung process is going on here."

I paused thoughtfully and then decided it was best to be blunt.

"I agree with you, Jared. That is the most likely scenario. But does that mean we kill him—now? That's what we're talking about here."

I turned to June, and then Faith, before returning my gaze to Jared.

"Does it mean he has no quality of life if he's in a rehab center? If he can't play golf or chop wood? Maybe reading history, following current events, speaking with family and friends are enough for him. He's a tough and determined fighter. He's wanted to try everything to get better, even when the chances were slim or the side effects significant. That's what he clearly told me."

"I've been beside him for thirty-six years," June answered gently. "After such a long time, you begin to share a person's mind. Peter wouldn't abide being in a chair, in some rehab facility, on oxygen, unable to move. He's a man whose whole life has been spent moving—around the globe and back."

A heavy gloom descended over the room. Faith began to cry. Carolyn looked uncertainly at me.

The problem is, I really don't know more than when James Hunt called. He is still a clinical enigma.

"Let's ask him," Jared asserted. "Go into his room. Explain what's happened. Tell him what we're discussing. Ask him what he thinks."

I recoiled. Tell Peter his son Jared is demanding to immediately disconnect him?

Peter Emery had been awake all night, undergone intubation, received numerous medications, including sedatives. What shape was he in to hear, listen, understand, and choose?

"His wishes were expressed to me once, on a visit when June didn't accompany him."

I recounted that Peter wanted no extraordinary measures, that if he was in great pain and there was no real chance of reversing the underlying condition, he wanted comfort measures only, without any undue prolongation of life. And if he became neurologically impaired, without cognitive function, so he was no longer "Peter Emery," as he put it, then he also wanted nothing done to extend his life.

"But are we there now?" I asked.

I had never faced a situation quite like this.

End-of-life decisions were familiar territory after caring for so many people with incurable blood diseases, cancer, or AIDS. Most of my patients had so-called advance directives documents, like living wills, about how much should be done to sustain them in the face of a terminal condition. I also sometimes encounter patients who ask, beyond their written words, to be "put to sleep." I tell them that the Dr. Kevorkian approach is not mine. Death-on-demand is, I explained, a perilous contract, since that request might be made out of depression, or fear, or the belief that pain and suffering will not abate. With modern medications, particularly long-acting opiates, and the expertise of hospice nurses to administer them and comfort the patient through the last days, the apparent need for Dr. Kevorkian disappears, since pain can be controlled and fear and despair addressed.

Everyone I care for desires a "good death," one marked by "dignity," meaning control, and having time for reconciliation with family, friends, self, and, among believers, God. But rarely have I witnessed such a smooth parting. There are usually unresolved emotions and festering conflicts, and the patient's residual anger and frustration at the lack of a cure. I had no diagnosis to explain the impending death.

It was true what Jared had asserted and June seconded. Peter's likely future—if he was to have any future at all—was bleak. The biopsy showed necrosis—widespread tissue death—and it was spreading now through both lungs. This meant he would have permanent respiratory damage, restricted to a chair, perhaps able to take a few steps. And, as I had told Peter and his family, there were no good treatments for the underlying condition, the myelofibrosis.

I reminded myself, in the long silence that followed my rhetorical question "But are we there now?" that I am never completely sanguine about exactly when to withdraw support from a patient. There is always that lingering sense that there might be something more to do. It could be illusory, something that reveals more about my doctor's reflex to fight disease than about clinical reality.

But ultimately I had to make that decision, in concert with the nurses, the family, and the patient, and despite the resistance to being thrust into the role of God, would act. That act was a passive one, but it still meant a sure death. I did not disconnect the respirator, as Jared had requested, or administer a lethal dose of morphine, and kill the patient within minutes. Confident that there was no pain on the prescribed analgesics, I simply refrained from doing more—no added antibiotics, no added pressors, no added inspiratory pressure on the respirator—and allowed "nature to take its course." Death was then gently passed back into the hands of God, who determined the exact when and how.

"Can we consider giving the amphotericin more time?" I finally offered to the Emerys. "If it is a fungus, the pneumonia might begin to improve. He's only been on the drug two days."

Carolyn caught my gaze, and I indicated she should speak.

"I agree with Dr. Groopman. I take care of people this ill all the time, and some improve, against all our expectations."

"It seems pretty remote," Jared replied. "Even if he survives, what will Dad be like? You don't know him like we do."

"I think I know him pretty well," I said. "Certainly, not like you do, but I have a good sense of what kind of man he was."

I heard the past tense in my reply and saw it as the first sign of meeting Jared halfway.

"Jerry's plan would give Matthew and Peter Jr. time to see Dad before he died," June said. The other two sons were still en route to Boston.

"I'm comfortable with that. If after forty-eight hours there is no improvement, we stop," Faith stated.

"What if there's a downturn before?" Jared asked. "Will you increase the pressors if his blood pressure drops again? Will you dialyze him if his kidneys fail? Will you shock him if his heart goes into an arrhythmia?"

He was, of course, right. Jared's special knowledge as an ICU nurse was coming to the fore. Not every contingency was encompassed by my forty-eight-hour compromise. There were still difficult decisions about what to do, and how much, even over such a short time span. While it was unusual to have so much clinical detail nailed down in advance, Jared knew such lack of definition could lead to scenes of confusion at the bedside, the nurses and interns intervening in ways that were not desired by the family or the patient.

"What do you think?" I asked Jared.

He hesitated, and then slowly articulated his position. "I guess we should give him the forty-eight hours. And as best we can, short of doing something inappropriate, support him." That meant, he specified, pressors for his circulation and electrical control of cardiac arrhythmias. But he drew the line at dialysis. "Now you can talk in detail with my father."

"Do you want to witness it?" I asked Jared. I didn't want to seem rude, but I wasn't sure if he trusted me.

He shook his head firmly no.

Carolyn walked with me to Peter's room. "That was heavy," she said.

I asked if she was okay with what had been decided.

"I think so."

Peter's eyes were half closed. His hair was matted with sweat, and the blue hospital gown clung to his wet chest. He was turned to face the respirator, the tube in his mouth at an angle to reduce its pressure against his trachea. The whoosh of the machine mixed with my soft words.

"Peter, it's Jerry."

I rubbed firmly over his sternum, imparting enough stimulation to rouse him. He slowly opened his glazed eyes.

"It's Dr. Groopman. It's Jerry. June and Jared and Faith are in the next room."

He nodded and tried to smile, but the gesture was distorted by the tube.

"Are you in any pain?"

He briefly closed his eyes, and then shook his head to say no.

"You couldn't get enough oxygen. We had to put you on the respirator. Your blood pressure dropped. We have you on medicines to support it. But we don't know if all this will be reversed. We'll make sure you don't suffer."

He closed his eyes for a long moment.

It was at this point that I would ordinarily stop. The unsaid was heard in the patient's mind. But I had promised to present the exact plan.

"I respect your wishes—what you told me you wanted when we spoke before, not to be sustained if you became vegetative, or if there was no hope and only pain. I've gone over that with June, Faith, and Jared. We're all in agreement. But we believe that even if we pull you through this, you'd be severely limited—largely confined to bed or a chair."

I paused to assess him. His eyes were open, he seemed attentive, but it still was hard to know how much was understood.

"We'd like to give the amphotericin another two days to see if it works—in case there is a fungus we couldn't find. But after that . . ." I paused and then forced myself to say it. "After that, we'll stop."

Peter's eyes shot open, his face registering shock and fear.

A burst of sharp pain filled my chest. I clenched my teeth and felt the tears collect in the corners of my eyes.

Slowly the terror receded from Peter's visage. His eyes lowered. He then looked back at me, his gaze plaintive, his brow raised in uncertainty.

"James Hunt is on the line," Carolyn called from the portal of the room.

June explained that she had called his home in the middle of the night, when Peter was deteriorating, and then realized James was away, at a wedding in New Jersey. She left a voice mail in case he checked his messages.

I stood off to the side, hearing only her yeses and nos in response to James's words.

"He wants to talk to you," June said.

I took the phone. James's voice was peppered by static.

"On a bloody car phone. Driving back up from New Jersey. A wedding I couldn't say no to. Can you hear me?"

I said he was clear enough.

"What the hell is happening? June said they want to let him go now. Seems awfully prescient, don't you think, old man?"

I said that Peter was "in extremis," medical jargon for a life-threatening state. I had affirmed to the family there was no treatment left for his myelofibrosis. June and the children believed Peter would not want to live as a respiratory cripple if he pulled through.

"June said Jared had you go into the room and tell that to Peter. Can Peter even bloody understand what's going on?"

I couldn't be sure but thought he had. I then detailed the compromise. We would support him for forty-eight hours, short of dialysis, and then stop intensive support.

"That gives us some time. It's too early to call it quits. You just can't disconnect the respirator. He'll die in minutes. Give him a chance and see if he rallies. If he doesn't, then his body is telling us 'enough.'"

I drove slowly back to Brookline, my mind crowded with the words and images of the morning. I concentrated hard on the streets, the changing stop lights and crossing pedestrians, anxious to get home without an accident.

The kids were seated around the kitchen table. Emily was drawing, having just learned how to construct faces with smiles or with frowns. She greeted me with a loud "Hello, Daddy." Michael was studying the ads in the Sunday *Globe*, hunting for the best discounts on video games. He didn't pick up his head from the paper. Steve was drinking a tall glass of chocolate milk, just back from his Tae Kwon Do class.

"You look terrible," Pam said.

I recounted the events of the morning.

Her face took on a grim look. "It's always so hard to know. But I agree that you can't just disconnect him. Are you going in over the holiday?"

The Jewish New Year, Rosh Hashanah, began that evening. It extended over two days. We were having dinner the first night with Pam's parents at our house and then the second night at the home of Russian friends, onetime refuseniks. Traditionally, each morning was spent in synagogue, engaged in prayer and contemplation, and usual work was not done in the afternoons. But all customs and religious restrictions were suspended if a person's life was at stake.

I told the Emerys I would come in at any time if needed, and no one should hesitate to call me. They insisted that the holiday not be disturbed unless it was absolutely necessary, and since we had agreed to forty-eight hours of support, I would be contacted then.

It was hard to detach myself, at dinner, in synagogue, at home in the afternoon. I kept trawling over the situation—the clinical details of the case, the interchange in the conference room, the meaning of Peter's terror followed by his uncertain grimace.

There were no calls from the ICU, although I checked the answering machine when we returned from synagogue. As soon as the sun set and the stars came out, the holiday then ended, I telephoned.

"He died four hours ago," Carolyn said.

"Why wasn't I called?" I asked with anguish.

She explained the Emerys absolutely refused to have Rosh Hashanah interrupted. June left her home number, so I could contact her after sundown.

"It went very smoothly," Carolyn continued. "He was awake on

Sunday and Monday. The two other sons, Matthew and Peter Jr., made it down. I think Peter Jr. was a little shocked by the plan, but he came around. By today Mr. Emery was pretty deep in coma. His kidneys were failing. We didn't have to disconnect him. He went on his own. Faith was holding his hand when he died."

I hung up and closed my eyes tightly. I saw Peter the first time we met, striding across the clinic waiting room, dressed in a brown tweed Brooks Brothers sports coat with a gold silk handkerchief flowing out of the breast pocket. His smile, broad and intelligent and inviting, lingered in the darkness of my mind.

"He died," I said to Pam.

"I'm sorry," she said as she opened her arms.

After wiping away more tears than I expected, I dialed the number in New Hampshire.

June's voice was tired, and I could hear her working to maintain control. "But it was beautiful, these last two days. As difficult as it was. We were all together again. We said things that needed to be said. We all told him how much we loved him, each in our own way. You gave us that opportunity, by negotiating the compromise."

I didn't know how to respond.

"We can't thank you enough for all you did. We requested an autopsy. Hopefully you'll learn what was going on in his lungs and use that to help other people."

⁓

The autopsy room was at the far end of the basement corridor of the old hospital building. The pathologist was gowned with two sets of yellow latex gloves and wore a broad Lucite shield over his face, to protect against splash exposures from potentially infectious fluids.

Peter's long body just fit onto the stainless-steel table, his toes touching the end rim. My eyes moved up along the line of his form, his sturdy legs, broad pelvis, and flat abdomen, stopping at his expansive chest. I stopped there for a long moment. Viewing the face and hands was always the most unsettling for me. These held the most character: the face, of course, for its remembered expressions; the hands, because they were what your flesh first touched in a welcoming grasp. In medical school, the very first week, we began anatomy by meeting our cadaver. I recalled how the hands and face were wrapped in a thick brown gauze, like a mummy; only after we were able to cut into dead human flesh without a second thought, some two months into the exercise, were these parts unwrapped. It still came as a shock, the first time I looked into my cadaver's face, and saw the inescapable reflection of my own being. And that jarring moment was relived each time I attended an autopsy.

"Look at the scale!" the pathologist exclaimed. The two excised lungs had jerked the spring so forcefully that the needle made more than three full circuits around the zero. The scale was like one in an old butcher shop, and indeed the lungs resembled liver, red-purple and very dense, rather than the normal light pink bellows. Peter's lungs weighed in at almost six kilograms, some thirteen-plus pounds, four times the normal weight for a man of his size.

The pathologist removed the lungs from the scale and placed them on a dissecting table. He sliced sharply into one; its beefy interior was exposed. He removed a wedge of consolidated red tissue, and I saw it was pocked by large black craters of necrosis.

"Hardly any airspace left," he concluded.

"Were all the special studies sent as I requested?" I pointedly asked.

The pathologist read back the list: tests for fungi, mycobacteria, and other microbes; electron microscopy in search of inflammatory complexes.

"His fevers and drops in blood pressure were an enigma to the end," I added by way of explanation.

"It's a wonder you could get any oxygen through, even on the respirator, with these lungs," the pathologist observed. "I'll call you in a few days with the results."

I waited each day for his call. Despite the family's sense that it was right to let go, and the pathologist's assertion that Peter's lungs were irreparably damaged, I kept playing the sequence of clinical events over in my mind. What if I had done the lung biopsy sooner? Would the vicious cycle of inflammation have been broken by increasing the doses of steroids started by James Hunt in New Hampshire? Or would we have found the elusive microbe in an earlier biopsy, before it was obscured by the necrotic debris of destroyed lung tissue? Should the amphotericin have been started as soon as Peter arrived, empirically, and despite the risk of kidney toxicity? If the autopsy revealed a fungus, would that earlier intervention have saved him?

The autopsy findings were outlined by the pathologist in a brief telephone conversation on a late Thursday afternoon; the typed report would follow Friday morning. I jotted a note, faxed it to James, and then called June.

"So there was no fungus, and the problem in his lungs seemed to come from the myelofibrosis," June said. She paused. "I'll tell the children."

I hung up the phone and exhaled in relief. Although I understood that decisions about life and death had to be made at times without all the answers, that truth was of little comfort. June and the children, somehow, had come to peace with Peter's passing before he died. Only now could I.

A Routine Case of Asthma

Just after eight o'clock one morning in November 1996, I was paged and asked to call Marianna Montero at our emergency room. I had not heard from Marianna since her father, Eduardo, had died of lymphoma seven months earlier. After Eduardo's death, Marianna and her mother, Isabella, had left Boston for Hyannis to live with extended family. Isabella had gotten a job cleaning in a bank at night. Marianna was a sophomore at the local high school.

"ER, John Healy speaking," a young male voice said crisply.

"This is Jerry Groopman, I'm returning a call from Marianna Montero."

"She's in the visitors' area, Dr. Groopman. I'll get her."

Some vanilla Muzak played on the line.

"Dr. Groopman, thank God, thank God," Marianna exclaimed. "Mama couldn't breathe again. I took her twice to the clinic at Cape Cod Care in Hyannis. They said it was just a routine case of asthma. Last night she couldn't see from one eye, and her arm was weak. I had to bring her here. I'm scared she's going to die. And I didn't . . ."

Her words dissolved in a sea of sobbing.

"Marianna," I said firmly, "whatever it is, we will help. I'll be right over."

I put aside the data I had begun to review for my morning lab meeting and headed for the ER. It was a chilly autumn day, a north wind swirling fallen brown leaves from the sidewalk. Traffic was already bumper-to-bumper on Brookline Avenue, the main route to the Harvard Medical area. I was relieved I had come in early and not been paged while stuck in my car.

Isabella was sitting bolt upright in bed. Her rich black hair was wet and matted, and droplets of perspiration circled her brow. There was a bluish tinge to her lips, and I noted how, with each breath, the sinewy strap muscles of her neck retracted in forceful upward arcs. Marianna was seated at her side, clutching the metal rail of the hospital bed.

"*Hola*, Isabella," I said loudly above the din of the beeping cardiac monitor and the gurgling oxygen flowing from a tank next to her bed. I gripped her moist left hand and gave it a squeeze. Isabella answered with a wan smile under the oxygen mask covering her nose and mouth.

Marianna stood up from her chair. She had grown since I last saw her in the spring. Tall and lean, she had her mother's jet-black hair and sculpted cheeks. Her dark eyes were wide with fear.

We embraced, and I told her I was glad she had brought Isabella to Boston. In a calm, deliberate tone, I explained I needed to review the ER notes and talk with the medical staff and would then return to discuss what had been found and should be done.

Isabella's triage form was on a clipboard at the nursing station. I focused on the salient details.

> 36 year old Hispanic woman . . . acute shortness of breath this evening . . . difficulty in seeing out of right eye . . . weakness in right hand and arm. Non-smoker . . . no prior history of lung disease. . . . Two months ago developed wheezing . . . told she had asthma by HMO doctor at Cape Cod Care. . . . Ventolin inhaler given . . . without benefit. . . . Last week started aminophylline and eryth-

romycin . . . no improvement in symptoms. . . . Respiratory rate 28 . . . temperature 100.3 F . . . arterial oxygen 53 . . . carbon dioxide 47 . . . chest X ray: diffuse interstitial thickening.

It certainly wasn't a routine case of asthma. First, it was unusual to develop the disease at Isabella's age. Most cases begin in childhood. Moreover, the Ventolin inhaler and the aminophylline, standard treatments that relieve it by opening airways, and erythromycin, an antibiotic to treat a complicating infection, were ineffective.

Isabella was on the brink of respiratory failure. Her respiratory rate was nearly twice normal. The level of oxygen in her arterial blood, normally around 98, was 53, precariously low. If this was not corrected by the oxygen mask, her organs would not function for long. Equally worrisome was the level of carbon dioxide. Carbon dioxide is the toxic gas generated by our metabolism and exhaled during respiration. Normal arterial carbon dioxide is around 40. The elevated level of 47 signaled a serious problem with her ventilation. Further buildup of carbon dioxide would acidify Isabella's blood, dull her brain, and become potentially fatal.

The chest X ray revealed diffuse interstitial thickening. The interstitium is the weblike lattice of tissue that supports the lung. On one side of this lattice are the air sacs, or alveoli; on the other side are the small blood vessels, or capillaries. Some process had thickened the interstitium throughout Isabella's lungs, thereby blocking her vital exchange of oxygen and carbon dioxide. Asthma per se did not cause these changes in the lung tissue.

I then read the findings from Isabella's physical examination. She had lost vision in her right eye. Her retina, viewed through an ophthalmoscope, showed large tracks of fresh blood marching along its dilated veins. On the roof of her mouth was a fan of blood-tinged spots, called petechiae. Fine crackling sounds were heard when she inhaled and exhaled deeply. Neurological testing confirmed weakness of her right hand and arm. I couldn't immediately think of a single diagnosis to explain all that was happening.

"It was just . . . the last . . . few months I had trouble breathing," Isabella affirmed between gasps. I instructed her to answer slowly, as best she could, so as not to worsen her air hunger.

The shortness of breath started at work. Her job involved vacuuming the offices in the main bank branch in Hyannis. To reach the corners, she had to move large furniture. She also cleaned the heavily used restrooms, scrubbing the soiled toilets and mopping scuffed tile floors.

"I went with her each time to the clinic at Cape Cod Care," Marianna interjected. "I told the nurses that for Mama to complain something had to be really wrong."

Dr. Matthew Sperry was Isabella's doctor at the HMO. He performed the employment physical when she started at the bank. She expected to see him at the clinic when she had trouble breathing, but he had been too busy. Instead, a friendly but harried nurse attended to her problem.

The nurse heard wheezing when she listened to Isabella's lungs. Maybe the dust in the bank, or the cleaning fluids, had triggered an allergy, the nurse said.

Dr. Sperry was informed of the findings, and the Ventolin inhaler was prescribed. Isabella used it as instructed when she felt her chest tighten and her breathing strained, but it offered little benefit. Isabella told the nurse she was hardly able to do her chores at the bank. Each sunrise she returned to her apartment and collapsed in exhaustion.

"I had to call many times to get Mama seen again," Marianna sharply recounted. "Finally, they gave us another appointment."

On that second visit, Marianna insisted her mother be examined by a doctor. They had to wait until the last appointment of the day was over.

Dr. Sperry listened to Isabella's chest and confirmed there were wheezes. He said she had asthma and wrote a prescription for aminophylline, to be taken orally three times a day, along with the antibiotic erythromycin. He said Isabella might have developed bronchitis, which worsened her asthma. Dr. Sperry cautioned both drugs could make Isabella nauseated but was sure they would "do the trick."

Isabella faithfully took the medications, although they did make her queasy. But her symptoms still did not improve. I asked if a sputum examination, blood tests, or chest X ray were performed at either visit. Marianna said she was certain none of these had been done.

Isabella began to cough in harsh spasms. Her lips turned a deeper blue and the cardiac monitor showed that her heart rate had jumped to 135. As the spasms gradually subsided, Marianna stared at me with mute terror.

Before I had a chance to examine Isabella, Rob Salerno, director of the ER, stepped into the doorway of the room. He beckoned me to the corridor.

"She won't be able to keep this up," he stated bluntly. "She's lucky she made it in from Hyannis. I just got her second set of blood gases back. The face mask isn't holding her. She needs to be intubated and placed on a respirator."

Rob Salerno is an excellent ER doctor. He is expert in the technical details of critical care medicine: the nuances of simultaneously administering multiple potent drugs, each with potential side effects, to resuscitate a patient in shock; the titration of intravenous fluids to fill a failing heart with needed volume without overloading the lungs or taxing the kidneys. And, equally important, he has a seasoned sense of who is going to precipitously deteriorate and how to intervene before the situation spins out of control.

Rob Salerno left to make the arrangements for the intubation as I returned to the Monteros.

I explained the necessity of the machine, saying I hoped it would be needed only for a few days, until Isabella's lungs improved and she could breathe on her own.

"Dr. Groopman?"

I turned to the voice. Lori McLaughlin, one of the senior medical residents, was standing behind me, a sheaf of papers in her hand.

"May I speak with you privately?"

I nodded and told Isabella and Marianna I would return shortly. I followed Lori to the end of the corridor, out of earshot.

"Here's your diagnosis," she gravely said.

Lori held out a thin rectangular slip of paper. Isabella's blood counts were printed in light blue numerals. My eyes fixed on "white blood cells: 220,000." Below the counts was a handwritten inscription: "98% blasts."

Isabella Montero did not have asthma. She had acute leukemia, a cancer of the white blood cells.

Isabella's disparate problems instantly merged in my mind, the shortness of breath and wheezing, the bleeding in her retina and mouth, the weakness of her arm and hand, the interstitial thickening on her chest X ray.

The leukemic white cells, called "blasts," do not grow, age, and die in a normal fashion. Rather, they proliferate wildly and fail to mature. Isabella had 220,000 white cells per microliter of her blood and nearly all blasts; normal is about 5,000, with, of course, no blasts. At this greatly elevated level, the hordes of cancerous white cells filled the circulation. The medical term was "leukostasis," or stalled white cells jamming the capillaries of Isabella's lungs like the cars backed up in front of the hospital. Oxygen could not pass in and carbon dioxide could not pass out through these clogged vessels. We had to clear a path quickly if we were to save her.

I then noted on the lab slip that Isabella's platelets, the blood cells that form a clot, were very low at 9,000. Normal is about 150,000. Leukemic blasts crowd out and suffocate the healthy cells in the bone marrow. This explained the fresh blood tracks in her retina, and given the findings of arm and hand weakness, a possible hemorrhage into her brain.

"Jerry, we need to intubate her now," Rob Salerno stated in an urgent voice. "Her last arterial oxygen was 47 and carbon dioxide 59."

I hurried with Lori McLaughlin into the room. Isabella's head was drooping, her eyes half closed, the falling blood oxygen and rising carbon dioxide dulling her brain. Rob Salerno quickly outlined to Marianna what he was going to do. Marianna said she knew about intubation from her father's case and began to cry

Rob Salerno flexed Isabella's head backward, forcing the front of her neck to arch forward and upward. Her jaw fell slack and her mouth opened. In a single deft move, he inserted a shining chrome instrument that displaced her tongue and lifted her epiglottis into view. With his other hand, he quickly passed a long, thin, lubricated plastic tube into Isabella's nostril and guided it down her throat and through her epiglottis, until it rested deeply in her larynx and into the trachea.

Isabella bucked at the invasion, her small cagelike chest jerking upward in a tense spasm. Dr. Salerno anticipated the reflex and elevated his hand in synchrony with her body. When she fell back on the bed, he peered down her throat. He announced that the tube was still in position. He then instructed a nurse to press air from a syringe attached to the side of the tube. This inflated a small balloon that cuffed the tube and fixed it in place in the central airway. Once the tube was declared secured, the nurse began pumping pure oxygen from a bag into the opening of the tube at Isabella's nostril.

Rob Salerno listened intently with his stethoscope to both sides of Isabella's chest. When he was confident that the tube was delivering air to her lungs and not into her esophagus and stomach, he signaled the nurse to remove the bag and attach a long, corrugated plastic hose from the respirator.

"Now let's get some platelets into her," I told Lori McLaughlin.

"We don't want her bleeding around the tube or extending the hemorrhage in her retina. I'm worried about her head. Get a CAT scan. And stat-page the pheresis team."

Pheresis was our best chance to rapidly reverse the leukostasis and unclog Isabella's vessels. The procedure involves filtering the blood through a special machine that separates the liquid part of the blood, called plasma, from the cellular part. We would return the plasma to Isabella while skimming off the hordes of leukemic white cells. Of course, this mechanical removal of the leukemia was only a temporary measure, since the cells would rapidly grow back from the diseased marrow. Chemotherapy would have to be started as soon as we finished the pheresis.

A team arrived to transport Isabella to the ICU. I told Marianna I would arrive there shortly.

I retired to the small doctor's office adjacent to the ER, exhausted. I glanced at my watch. It was 9:10 A.M. Less than an hour had passed since the page from Marianna. The events of the morning felt like a volley of merciless punches from a tenacious boxer.

I took a cup of overbrewed ER coffee and sat to collect my thoughts. Nothing more could be done for the moment. But I wanted a clearer picture of what had happened in Hyannis.

I dialed the number for Cape Cod Care to talk with Dr. Sperry. After two rings, an operator answered and immediately asked me to hold. Handel's "Water Music" played. I followed the sweeping second hand on the wall clock. As each minute passed, a recorded message intoned: "Please continue to hold. All our operators are currently busy. Your call will be answered in the order it was received. Cape Cod Care: Celebrating Fifteen Years as the Cape's Premier Health Plan." After the third repetition of the message, I switched the phone to speaker.

My thoughts drifted to Eduardo, Isabella's late husband. A small, wiry man with an olive complexion and curly black hair and mustache, he became bald and pale from months of intensive chemotherapy for his lymphoma. Eduardo approached his cancer in a stoical and determined fashion. Despite the debilitating therapy, with its side effects of blistering mouth sores, diarrhea, and anemia, he stubbornly persisted in trying to continue as the family bread-winner, working when he could as a mason, the trade he learned in the Dominican Republic.

Eduardo's lymphoma transiently regressed with our treatments but then grew explosively. It rapidly filled his abdomen and encased his lungs.

When I told Eduardo that our treatment had failed, he closed his eyes and nodded knowingly. He then firmly instructed me not to place him on a respirator or otherwise mechanically prolong his life; he only wanted to be at home with his wife and daughter as long as possible, and to receive last rites. Eduardo died with Isabella and Marianna at his bedside in the amber light of an early April morn-ing.

"Yes, how can I help you?" the operator intoned after about six minutes.

"Dr. Matthew Sperry, please," I said. I switched the phone back from speaker.

"One moment, sir."

A series of clicks transferred me to a recorded message. I was in-formed that I had reached the office of Primary Care Associates. I was to press 1 if I had a Touch-Tone phone and hold if it was rotary; I pressed 1. I was then presented with a menu of options: press 1 for prescriptions, 2 for appointments, 3 for billing questions, 4 for test results, 5 if I needed approval to see a specialist, 6 for the nurse as-sistant, and 7 for a doctor. I pressed 7. A recording asked me to spell the doctor's name, using the keys on the Touch-Tone phone. For Q or Z, I was to press 0. I pressed 7 for S, 7 for P, 3 for E, 7 twice for the two Rs, and 9 for Y. After each letter, I paused to check the phone's face, fearful of making an error and being returned to the music.

"Dr. Matthew Sperry," a metallic, computer-generated voice intoned. "Press 1 to confirm your choice."

I did. A prerecorded message addressed me.

"This is Dr. Matthew Sperry. I am not currently available. Please leave your name and number at the sound of the beep and I will return your call as soon as possible. If this is a medical emergency or other urgent matter, please press zero now."

The voice sounded middle-aged and with a strong Maine accent, the *a* long and flat. I was irritated, wound up from the crush of events in the ER, and now unable to locate Isabella's doctor. For a moment I hesitated. Should I leave a message or press 0? Isabella was en route to the ICU, receiving concerted attention and care. It was no longer strictly a medical emergency. But I wanted to verify the details of her clinic visits and treatments, and be certain that nothing relevant to her medical condition was unknown to us. Dr. Sperry might not pick up a nonurgent message until late in the day. I pressed 0.

"Primary Care Associates," a live person announced after a brief pause.

"I'm Dr. Jerome Groopman, in Boston, at Harvard Medical School," I said, knowing that emphasizing *Doctor, Boston,* and *Harvard* would get Isabella's case more attention. "And I'm trying to reach Dr. Matthew Sperry on an urgent matter. To whom am I speaking?"

"This is Nancy, the clinic receptionist. Is it about a patient?" Her voice was pleasant.

"Yes. Isabella Montero."

"Are you one of her physicians?"

I explained that I had cared for her husband in Boston, adding he had died of lymphoma last year and Isabella was a single mother raising her teenage daughter in Hyannis. Mrs. Montero was admitted this morning to our ICU and I was calling for medical information. I gave the receptionist these details to recognize her as a fellow professional and to spark her sympathy so she might make an extra effort to help.

"Dr. Sperry is not in yet. But I'll pull Mrs. Montero's chart and find one of the nurses to speak with you."

I thanked her warmly. Soon an affable nurse named Julie was on the line. She read from the medical record.

Isabella was first seen in March 1996 when she was hired at the bank as part of its night shift cleaning staff. All bank employees were enrolled in Cape Cod Care; Julie explained it was part of their contract. Dr. Sperry had written at the time of employment that Isabella's medical history and physical examination were "unremarkable."

The first clinic appointment was on October 9. Carolyn, one of the HMO's nurses, had seen Isabella. The second visit was the afternoon of October 24 with Dr. Sperry. The diagnosis of asthma and the various prescriptions were all as Marianna had recounted them to me. Nowhere in the chart were there results of a sputum examination, blood test, or chest X ray.

I thanked her and asked to be called as soon as Dr. Sperry arrived.

I found Marianna in the solarium outside the ICU. She barely lifted her head to face me as we spoke.

"The second time, when Dr. Sperry saw Mama, I knew something was really, really wrong. But what could I do?"

I replied that Marianna shouldn't in any way blame herself—she was in no position to challenge the diagnosis of asthma and the treatment prescribed by Dr. Sperry.

"We need to focus on what's happening now and do everything possible to get your mother better."

Marianna did not reply. My words sounded hollow even to me, but discussing the misdiagnosis seemed to serve no useful purpose. We sat silently together a few minutes longer. Then I left, saying I'd return in the early afternoon.

My laboratory meeting began an hour late. I tried to detach from the Montero family and focus on the experimental data being presented, but it was impossible to compartmentalize my thoughts. Each aspect of the presented experiments—the concentrations of chemicals, the numbers of cells, the temperature of the mix, the time of incubation—reminded me how success in science requires both an obsessive attention to minutiae and a mind open to the unexpected. The smallest details, easily taken for granted or overlooked, so often trip you up, causing an experiment to fail. Even more dangerous than details are assumptions. Failure to reexamine your initial hypothesis when the experimental results do not fully support it leads you down fruitless paths.

Clinical medicine differs from laboratory science in form, not substance. The same restless vigilance is needed, so you'd better keep your eyes open and your nerves sharp, and never stop questioning your first impressions. Only with patients, unlike cells in a test tube, you may not have a chance to try again if you fail.

"I vaguely remember her," Dr. Sperry began. "I'm assigned everyone from the bank's four branches." His voice had the fatigued baritone of middle age, and his accent was unmistakably rooted in Maine. "I'm looking at her chart now. Nothing wrong when she started as a cleaning girl at the main office. She was wheezing, and I gave her Ventolin and then some aminophylline for presumed asthma. I added erythromycin for a possible bronchitis about a month ago."

"There wasn't a white count or sputum culture or chest X ray when you thought she had bronchitis?"

Hearing no reply from Dr. Sperry, I continued.

"We usually do that when symptoms don't respond to empiric therapy."

"Dr. Groopman, come down from your ivory tower and try working here in the trenches." His voice spat acid, and I recoiled in stunned silence. Rarely do physicians confront each other so directly. "How many patients do you see in a morning? Six, maybe seven? With residents and fellows to do your scut work.

"Spend a week here with me. I have ten Isabella Monteros a day in the waiting room, complaining in broken English they need time off from work because they're tired or can't breathe.

"We have proven guidelines for what tests to order and what treatments to give. It's not cost-effective to do more than I did for a routine case of asthma. She wasn't bringing up sputum. And a chest X ray and blood counts are outside our clinical algorithm for these cases. How many turn out to be a rare manifestation of leukemia and leukostasis? For every thousand it's asthma 999-plus times. So don't interrogate me."

His condescending tone, his failure to express sympathy for Isabella or regret at missing her diagnosis, brought me to the boiling point. I clenched my jaw to stop from answering back.

I forced myself to step aside from Dr. Sperry's attack and had to admit to one painful truth. I did occupy a privileged position in the world of modern American medicine. My salary was fixed, coming from research grants and an endowed chair at Harvard. If I saw one person or ten in the clinic, it made no difference. If I wanted to spend an hour rather than fifteen minutes to examine and talk with a patient, I could. No one imposed "cost-effective" clinical algorithms to stay my hand from pursuing a more intensive evaluation.

It was very different for full-time clinicians, particularly those in community HMOs. Over the past decade, with a nationwide shift to managed care, most doctors are forced to see ever more patients per unit hour and to be parsimonious in their care. It is always cost-effective to do less when outcomes are measured in the distorted mirror of the group. The art of medicine, rooted in the needs of the single patient and the judgment of the individual practitioner, is being dismissed by the architects of managed care as archaic and inefficient. The rules and regime of the factory prevail.

Blame for this monstrous creation is to be shared not only by insurers and HMOs but also by doctors. Under the old fee-for-service system, the more tests run and procedures performed, the more money the physician made. Some doctors gouged, ordering unnecessary tests, admitting patients to the hospital for problems that could be solved in the outpatient arena, even doing questionable surgery. The costs of health care skyrocketed. Companies and individuals buckled under the burden. Health-care expenditures fueled the decades of explosive inflation. In response to this, the cry went up to restrain spending and eliminate waste. But the pendulum is swinging too far in the other direction.

Responsible and committed physicians are increasingly frustrated and despairing. The joys of being a doctor, derived from the special intimacy between physician and patient, and from the challenge of the craft, are evaporating. Many doctors, in HMOs and out, are seeking other careers or defect to the other side—becoming administrators who dictate cost-effective algorithms at a safe distance from the daily demands of practice.

I was becoming calmer, glad I hadn't answered immediately and escalated the confrontation. It would be of no strategic benefit. Isabella was still a member of Cape Cod Care. If she survived this hospitalization, she would return to Hyannis needing follow-up care from Dr. Sperry. Moreover, I knew my hospital faced a negotiation with the HMO about reimbursement for her emergency admission. We were not part of the HMO's network, and it was unclear whether they would view Marianna's decision to bring her mother to Boston as acceptable.

"We can debate the complexities of health-care delivery another time," I evenly offered. "I'd like to have my hospital admitting office contact the person in your group who approves out-of-network admissions. Do you have a name and direct number to help move this along?"

"I do," he said, his voice only minimally warming. "I'm not sure how they'll handle the case. The daughter drove her up without calling our clinic's emergency number. All emergencies go to Cape Cod Hospital, and transfers out of network are arranged from there for special needs outside our capabilities."

Dr. Sperry gave me the administrator's name and number.

I now worried not only about the complex clinical management of Isabella's leukemia but whether the HMO would agree to pay for her care. The financial success of an HMO depends on keeping as many patients inside a closed network as possible, since costs within the network are relatively fixed. Isabella's hospitalization in our ICU would run tens of thousands, or possibly hundreds of thousands, of dollars. An HMO administrator who gave approval for this added expense had to have good reasons for his decision.

Isabella's CAT scan showed a small circumscribed hemorrhage in her left brain. This explained the right hand and arm weakness. The pheresis succeeded in lowering her blast count to 20,000 within a day. Transfusions raised her platelet level to prevent further bleeding. It would take time before we knew if there would be permanent damage to her brain and retina.

Treatment of acute leukemia is one of the most intensive therapies in clinical medicine. High doses of two drugs—ara C and daunorubicin—were needed to totally destroy Isabella's cancerous white cells. This toxic chemotherapy would also destroy nearly all of her normal marrow cells, as well as potentially damage her heart, liver, skin, bone, and gastrointestinal tract. The hope was that the leukemic cells would be more sensitive to the treatment than the healthy tissues, so the cancer would be eliminated and the normal cells, including those in the marrow, could regenerate. Waiting a week or more for that regeneration, Isabella would be without any white cells to defend against infection and without platelets to clot. Many patients die during that period from overwhelming infection or uncontrolled hemorrhage.

I explained the rationale and the risks of the three drugs to Isabella and Marianna. They were familiar with the scenario from Eduardo's lymphoma. And, like her husband, Isabella did not hesitate. Unable to speak because of the intubation, she affirmed her desire to go ahead by writing "Y-E-S" on a yellow legal pad. She then pointed to her mouth.

"How long before the tube comes out?" I translated.

She fixed her brown eyes on mine.

"Hard to say."

I elaborated that her acute deterioration was due to the white cells clogging the interstitium of her lungs. It was also possible, with her low-grade fever, that she had an infection. Each was being treated and would, we hoped, soon subside. We could then detach her from the respirator.

Isabella steadied the pencil in her hand as Marianna held the tablet of paper.

I watched the frail letters drawn as deliberately as a toddler's.

"I-F N-O-T?"

I looked gravely at Isabella. Marianna stood tensely at her side, still holding the clipboard.

I was tempted to evade her question and say "We'll cross that bridge if we need to" or some other cliché. It can be daunting for a patient in the midst of severe illness to know all possible negative outcomes. These, rather than the positive ones, become the dominant scenes played in one's mind. But I knew I should respond forthrightly. First, it was Isabella's right to know. And second, I sensed the question was a test, to see if I would communicate with the same unflinching honesty as I had with Eduardo.

"There are some people who have to be supported on a respirator for weeks, months, or longer, if the leukemia does not respond to our treatment, or if an infection cannot be controlled, or if their brain is impaired and they lose the drive to breathe."

The corners of Isabella's mouth tightened as she gagged around the tube in her throat. Her eyes closed in pain and then opened, wet with tears. She shook her head side to side and in large block letters rapidly scrawled: "N-O."

～⌒⌒◯

Dr. Sperry called on the eighth day of Isabella's hospitalization in our ICU.

"How's my patient doing up there?"

In a cool tone, I reviewed the pertinent data. There were still leukemic cells in Isabella's blood, but with the combination chemotherapy, they were sharply falling. Her hand and arm weakness and visual loss were unchanged. Her fevers had increased, and her respiratory function had deteriorated. We worried she might have a pneumonia in addition to the leukostasis, and she was receiving antibiotics. How long she would remain on the respirator was uncertain.

"Sounds rocky, but I guess that's how leukemia is."

I tersely agreed.

"Mrs. Montero's hospitalization won't be covered," Dr. Sperry continued. "The administrators here are tough. The daughter should have brought her to Cape Cod Hospital. Our oncologist, Barry Sonder, is young but a competent guy. He would have cared for her."

I paused. I could again feel anger rushing through my limbs. I was tempted to finally smash the façade of collegiality, but I still feared an outright blowup would endanger Isabella, if not now, then in the future. The HMO was still her only coverage. She would need its services if she survived.

I ended the conversation, telling Dr. Sperry I had work to attend to.

"Please keep me posted, Jerry. May I call you Jerry?"

I said yes.

Isabella was successfully weaned from the respirator at the end of the third week of her hospitalization. She proved to have a virulent bacterial pneumonia called pseudomonas; it eventually yielded to a combination of potent antibiotics. We also administered a protein called G-CSF that boosts blood cell growth, expediting the return of normal white cells.

The side effects of the chemotherapy were severe. Isabella's mouth blistered, and she developed diarrhea. We supported her with analgesics and intravenous fluids. For several days she was alternately confused and somnolent, a side effect of high doses of ara-C. But there was no further bleeding, and the strength in her hand and arm improved. Her vision was harder to assess.

By the beginning of the fourth week she was transferred out of the ICU to a regular medical floor. There, Isabella took some slow first steps and began to eat again. Her blood counts slowly returned to normal, but a repeat bone marrow examination revealed she still had leukemic blasts, although much reduced in number. She would need further combination therapy in an effort to secure a complete remission. About one third of patients enter remission with the first cycle of treatment, and about another 5 to 10 percent require a second cycle. Isabella's prognosis was poorer than most. The height of the white count and the type of blasts she had, called monoblasts, weighed against achieving a cure.

Isabella returned to Hyannis, frail and despondent. Marianna wanted to take time off from school to help her at home, but Isabella would not allow it. Extended family in the same building would help during the day when Marianna was in classes. Isabella's cousin Herman managed a local supermarket and would assure regular deliveries of food and household items.

"I hate going back to that clinic," Marianna said bitterly when I called to check on Isabella. "The nurses are okay, but Dr. Sperry . . ."

I really wasn't sure what he was talking about. He said Mama has such a rare disease and then went on and on about how tests are 'cost-ineffective' for wheezing."

Marianna paused and her voice tightened.

"Do we have to stay there? Can't her tests be someplace else? And what about her next treatment?"

I explained that Isabella was required to receive her care there. She might request another doctor. She was on medical leave from work but still had coverage and benefits. Isabella had refused to consider welfare. Medicaid, the state public assistance program, was really her only choice.

"Herman thinks she got sick from the job," Marianna continued. "I went to the Hyannis library and looked up acute leukemia. Toxic chemicals can cause it. There are all sorts of chemicals in the cleaning fluids at the bank."

I said it was difficult to be sure. Benzene and other volatile solvents were certainly linked to the disease. But these were mostly petrochemicals and present in paint remover and varnishes. Moreover, Isabella used gloves when she cleaned, making the exposure less likely. In most cases of acute leukemia, the cause cannot be identified.

The second cycle of chemotherapy was given by Dr. Barry Sonder in Hyannis. After four weeks of intensive therapy and support with transfusions and antibiotics, he called and informed me there were still leukemic blasts in Isabella's marrow. Her disease did not yield to the treatment.

Later that week, the Monteros drove to Boston to see me. Isabella looked skeletal, her complexion wan, her eyes dull and sunken. Marianna still held out hope for her mother. I set out what options remained. There were so-called phase I trials where new experimental drugs were given so as to ascertain their side effects. Isabella asked if the experimental drugs were likely to cure her. I answered that the chances were remote. At best, they might temporarily stay the disease. Moreover, since the experimental chemotherapy was expected to be toxic, it would require hospitalization and medical monitoring. But now Cape Cod Care would not be a stumbling block, since the experimental trials were subsidized by a federal grant and would cost the HMO nothing.

Marianna asked about marrow transplantation, and I explained that it rarely succeeded when the leukemia was completely resistant to chemotherapy. Moreover, we had typed Marianna and others in the family at the time of Isabella's first admission and no one matched as a donor.

"It is now up to God," Isabella whispered.

Marianna started to object, but Isabella shook her head gravely. She told her daughter that, like Eduardo, she recognized the end. Now was the time to put things in place so Marianna would be cared for. Herman was the head of the family. He would be responsible after she was gone.

I returned from the clinic to my office and lowered myself into the chair at my desk. For a long time I sat there, too enervated to move or speak. The right choice had been made, but that only seemed to intensify the pain.

———

"She was screwed over big time by the HMO," Mr. Ernesto Badilla brusquely asserted. "One more Hispanic cleaning lady they think they can ignore. But not with Ernesto Badilla as her lawyer. We're going after *them*."

Some two weeks had passed since my conversation with the Monteros. Isabella was weaker each day, tenuously kept alive by weekly transfusions and oral antibiotics.

Marianna's bitterness about Dr. Sperry's care increased as her mother moved closer to death. She could not stop blaming herself for not aggressively demanding that Isabella be seen by another doctor. I tried again to assuage her sense of guilt, reiterating that most adults, not to mention a high school sophomore, don't feel they are in a position to challenge medical professionals. She found cold comfort in this, so I was not surprised when Marianna called and informed me that Herman knew an attorney, who would contact me. I wasn't sure whether Isabella was aware of this step but didn't ask.

"I've been around the block more than once and know the tricks," Mr. Badilla continued. "I do personal injury and malpractice. And I do it well. Dr. Matthew Sperry didn't take good notes in his risk-management courses."

Risk-management courses are didactic programs offered to physicians to help them avoid malpractice suits. With the proliferation of such litigation, some of it justified and some not, hospitals and medical societies have begun to educate health-care providers about behaviors that engender lawsuits. In Massachusetts, all staff must attend presentations by lawyers, hospital administrators, and clinical psychologists who explain how and why doctors and nurses are placed in adversarial positions with a patient and his family.

I learned from these courses that most litigation grows not out of honest errors or even frank malpractice but from unresolved anger and poor communication. Physicians are not used to admitting when they are wrong and plainly stating to the patient and family that an error was made, a lab test overlooked, a finding missed on a physical exam, or an incorrect drug prescribed. In Isabella's case, Dr. Sperry raised the defense that it was not cost-effective to do blood counts and a chest X ray for a case of presumed asthma. As far as I knew, Dr. Sperry never expressed regret. What Marianna saw in his words was greater concern about his own image and propriety than about the fate of her dying mother. His obtuse cost-ineffective explanation only further inflamed her.

There is also, of course, a critical catalyst in the mix of events and personalities that can ignite litigation: the lawyer. Some, I believe, represent their clients appropriately and seek remuneration for damages due to negligence or egregious errors in diagnosis and treatment. Others are ambulance chasers who care only about a prospective fifty-fifty split between client and attorney in the case of a court's award.

"You're a friend of the family," Mr. Badilla continued, "and we know we can count on you in this case."

I replied I would report the events as I knew them to have occurred.

"A lot of doctors cover each other's butts," he added, his tone sharpening. "You know, the 'Don't shit in your own nest' kind of a thing."

I was wary of engaging further. I repeated that I would honestly recount the details of Isabella's care and then ended the conversation.

I sat thinking before making any subsequent phone calls. Although I was inclined to contact Marianna to verify what Mr. Badilla had said, I decided first to notify my hospital attorney, Dorothy Kleinman. Dorothy is a lawyer of remarkable organization and diligence who likes to know about such matters even if they never come to pass. She would caution me about what I should say to the Monteros; everything that happened from now on could later be made the grist of testimony. I would reassure her that my relationship with the family was a good one, and that the care Isabella had received at our institution was excellent. My opinion of how Isabella was treated at the HMO was another matter.

⌒〜〇

Dorothy Kleinman and I took a cab downtown for my deposition in the law offices of a prestigious Boston firm. The traffic on Storrow Drive was unusually heavy, an accident at Leverett Circle causing a backup. Cape Cod Care had gone outside its usual counsel to use the services of this elite firm, a fact that Dorothy took as an indication of the seriousness of the suit.

"Detach yourself from your feelings," she instructed me before the deposition. "Remember—no amplifying, no editorializing, no speculation. Yes or no answers whenever possible."

I had never been deposed in a malpractice case before. Although being an expert witness in malpractice or other medical cases was a lucrative opportunity that many academic physicians pursued, I disliked the idea of such work. It was not because I didn't want to shit in my own nest, as Mr. Badilla had crudely put it. Nor was it because I was opposed to patients and families receiving compensation for deficiencies in medical care. Rather, I was disturbed by how often right and wrong were obscured behind artful legal maneuvers. I didn't want to be a "hired gun."

Dorothy Kleinman also emphasized that the legal dimensions of this case did not extend to the broader problems of managed care in America. The suit was against Dr. Sperry. A law passed in 1974 made it difficult for patients to sue HMOs for denial of care. Dr. Sperry had followed the institutional clinical algorithm, and the HMO was not obligated by law to pay for Isabella's treatment in Boston. That made the family's case more difficult. Mr. Badilla had to prove damages due to the delay in diagnosis or treatment.

"There are damages, Dorothy," I asserted. "The cerebral and retinal hemorrhages might have been avoided if the leukemia had been detected earlier. Isabella never fully regained vision in her right eye. And the failure of the leukemia to be eradicated by the chemotherapy might be due to the massive burden of cancer, reflected in a blast count of over 200,000. If Isabella had been diagnosed earlier, when there was less leukemia in her system, we might have saved her."

" 'Might,' 'if' . . . all speculative," Dorothy countered, playing the devil's advocate. "Furthermore, the visual problem is minor and wouldn't significantly limit her. The argument about cure would be contested by expert witnesses defending Cape Cod Care. They'll say the odds were so long against her anyway."

"I don't give a damn about expert witnesses," I bitterly replied.

"You fight like hell and don't give cancer an inch when you do oncology. Sure, the chances of cure were strongly against Isabella. But they weren't zero, and Sperry made them worse."

Dorothy looked plaintively at me, nodded, and then reminded me to keep my feelings to myself.

We sat in high-backed leather chairs at a teak conference table. Antique prints of American schooners and steamships lined the walls. A stenographer was situated discreetly in the corner. The two lawyers representing Cape Cod Care sat opposite Dorothy and me. One was in his mid-fifties, portly with a pasty complexion offset by a red silk bow tie and starched pink shirt. His associate was a younger woman, probably in her late thirties, lean and dressed in a dark business suit. She gave me a severe smile.

After I affirmed that I was indeed who I was, the associate asked if I was expert in asthma or lung diseases.

The older attorney assumed a mien of profound interest.

"As a specialist in cancer, blood diseases, and AIDS, I care for many people with pulmonary complications of their underlying disease or its treatment, and—"

Dorothy Kleinman turned sharply to me and glared.

I stopped speaking. I looked back at Dorothy beseechingly. Her gaze increased in intensity.

"No, I am not a specialist in asthma or lung diseases."

"Then we shall not ask you any questions about Mrs. Montero's evaluation for wheezing on the two occasions in October when she was seen at Cape Cod Care."

I opened my mouth to speak, but Dorothy Kleinman subtly signaled silence. I wanted to tell them that no patient should be dismissed *twice* as having asthma and given an inhaler and an antibiotic without examining her sputum, obtaining a chest X ray, and checking her blood counts. I didn't care what clinical algorithm existed at the HMO, or whether it was cost-effective or not when analyzed against more than a thousand cases of wheezing. That was not clinical medicine. That was robotics. Medicine meant thinking about the patient as an individual and keeping your mind open to the unusual and unexpected. But I kept silent.

The older attorney picked up where his associate had ended. "Dr. Groopman, you stated you are a specialist in hematology and oncology, correct?"

"Correct."

"And do you have any reason to doubt the competency of Dr. Barry Sonder, the board-certified hematologist-oncologist at Cape Cod Care?"

I paused. I had no reason to question Dr. Sonder's competency. But that was not the point. "Would you choose to be in Cape Cod Care if you had acute leukemia, with doctors you don't know, who have no history with your family?" I asked the lawyer sharply. "Even the best community clinics treat only a handful of cases of acute leukemia each year. I bet you'd be in a university hospital in Boston, with the best specialists on a dedicated ward with specialized nursing."

"Are you saying Dr. Sonder is not competent?" he immediately fired back.

Dorothy Kleinman interrupted. She said she needed to speak to me privately. I reluctantly stood from the table and retired with her to a small room adjacent to the one we had been in.

"Jerry," she began with the exasperated tone of a grade school teacher addressing an errant pupil, "it serves Mrs. Montero no good, and you do yourself no good, debating these lawyers. You have to keep yourself under control."

"But I didn't say anything damaging. Dr. Sonder is competent. But he may see at most three or four cases of acute leukemia a year. And I believe the Monteros should have had the right to seek care in Boston for medical and emotional reasons."

"Jerry, for the nth time, what you believe has nothing to do with their questions." Dorothy paused and her tone softened.

"Be grateful this will be your only time spent with lawyers. Badilla won't have to depose you. The case will never come to trial."

She had already spoken with Ernesto Badilla about his case. He faced a number of problems. It would be difficult to ascribe the leukemia to exposure to toxic cleaning fluids at the bank, as Isabella's cousin Herman had suggested. And since Dr. Sperry had stayed within the guidelines of his HMO, there was no overt malpractice. There also was no basis for Isabella to say she suffered by being forced to see Dr. Sonder. Badilla would argue as I had, that Isabella had suffered due to the delay in diagnosis and had some loss of sight. But that wasn't his trump card.

Dorothy's eyes narrowed and her voice grew low, like a conspirator passing on a secret code.

"It's the PR that gives Badilla the upper hand over Cape Cod Care. The *Globe* would have a field day with this. 'Dedicated Hispanic widow refuses welfare, slaves away cleaning at a bank, misdiagnosed by the fastest-growing HMO on the Cape. Daughter drives in the middle of the night to be rescued in Boston.' The two lawyers here know there isn't a jury in the Commonwealth where most members aren't pissed off at HMOs, or know some horror story about managed care. Now, let's go back into the conference room and give them dull yes and no answers and wrap this up."

Less than a month later, Dorothy Kleinman's words proved true. She called to inform me that a settlement had been reached. She didn't know the sum, and I, under no circumstances, should ask the Monteros. The case was sealed as part of the agreement to settle. There were to be no public disclosures of either the incident or the remuneration. No guilt was assigned to any party, and no apology was offered to the Monteros. Dorothy added that it was likely that Ernesto Badilla would receive half or more of the payment.

I wondered what impact this would have on Dr. Sperry and the workings of the HMO.

"Probably minimal," Dorothy said tiredly. "Hopefully, he'll be more thoughtful. They'll ramp up their risk-management training and discuss with the other doctors whether an apology to Marianna might have averted the suit. But you know the name of the game these days: move 'em in, move 'em out."

I drove to Hyannis on a slate gray February morning for Isabella's funeral. The clusters of naked trees lining Route 3 to the Cape stood like stripped bones in an ossuary. Bold clouds threatened rain, and I imagined the frost-covered cemetery ground giving way to viscous mud.

Isabella had stayed at home with a hospice nurse administering morphine to control the pain. Marianna called me shortly after her death. Although it was fully expected, her woeful voice cut into me. I told Marianna again how deeply I felt about her mother, and how I wished there had been more to do for Isabella. Marianna thanked me and assured me that her cousin Herman and his family were looking after her.

There was little traffic, and as I drove, I thought again about how the world of medicine was changing. So long as you were healthy, managed care seemed fine. You parted with less each month from your paycheck for coverage, and your company was made more profitable by reducing overhead. But when you fell ill, seriously ill, you found yourself at the mercy of a marketplace where the bottom line is measured in dollars, not quality or compassionate care.

There is the beginning of a backlash. The restriction of a patient's choice of a personal doctor and the ratcheting down of clinical services clash with the American belief in health care as a right. Patients and their families are objecting to the deficiencies and indignities in services. Doctors and nurses laboring in the system are increasingly unhappy. Together with patients, they are lobbying for a patient's bill of rights and loosening of managed care restrictions. How medicine will evolve cannot yet be envisioned. But there is a sharpening awareness that there should be no compromise between what patients deserve and what doctors and nurses provide.

I crossed the Sagamore Bridge over the Cape Cod Canal and entered the rotary onto the Cape itself. I was still haunted by the sense that if Isabella's diagnosis had been made sooner, she might have survived. I played out in my mind that wishful scenario. Intensive chemotherapy was given at the very beginning of October. She had a difficult course, as happens with all cases of acute leukemia. There were weeks of high fever and aggressive infections. She required multiple antibiotics and transfusions, to support her until the destructive effects of the chemotherapy resolved. Then came the revealing moment when her bone marrow again was sampled. I saw in my mind's eye a microscopic sea of diverse marrow cells, each maturing and content, without a trace of deadly blasts.

It was a short-lived daydream. I turned south toward Hyannis and readied myself for Marianna's tears, and for my own.

From the "Old School"

Sarah Beckwith, a friend since college days, called me at home Labor Day weekend in 1996. We had kept in regular touch over the years. I attended her wedding at the family estate in Beverly, on Boston's North Shore, and saw her on occasion for coffee or a quick lunch when she was in the Harvard Medical area. She was divorced now, without children, and worked as an investment analyst at a large downtown firm.

"I'm really sorry to bother you on the holiday, but it's my father," she said. "He's lost weight and has had a fever for weeks. Daddy is the kind who doesn't say much. He's from the 'old school.' He saw Dr. Hugh Bisson, who's been our family physician forever. You may know him from your training at Mass General."

I said I knew him only vaguely.

"He's semiretired now. Today, Daddy said it's some kind of lymphoma. Dr. Bisson referred him to a specialist at Essex Hospital, Dr. Andrew McBride.

"I went on the Internet," Sarah went on. "There are so many different kinds of lymphoma. I called Hugh Bisson and asked for details."

She paused gravely.

"I didn't like the way he talked to me—like I was a little girl. He said, 'Leave the worrying to me.' And that it was too technical to explain to a layman exactly what kind of lymphoma Daddy had. Just that it grew slowly and the treatment was simple."

Dr. Bisson's response concerned me, but I waited to hear Sarah out.

"Could you contact them and find out what's going on? I apologize again for calling today, but I'm unnerved by the whole situation."

I assured Sarah she need not apologize. I searched my memory from years ago. Hugh Bisson occasionally referred patients from the North Shore to the hospital, but I hadn't been directly involved as an intern in any of his cases. I recalled him as a tall, gaunt, pie-faced man with a shock of thick white hair and a lumbering gait. I thought he must now be in his seventies.

"Sarah, I can't call either doctor cold. The protocol is that your father informs Dr. Bisson and Dr. McBride that he wants them to discuss his case with me. Your father also has to give me permission to relay what I learn to you." I elaborated that this procedure recognized Bisson and McBride as the primary caregivers and my specifying it was a courtesy call would avoid any misperception that I was interfering in their case.

"I'll have to talk with Daddy. He's a stubborn, difficult man, and since my mother's death, he's only gotten worse."

Sarah's mother, Claire, had died some six years earlier. The family was vacationing in the Adirondacks and her mother, a vigorous sportswoman, was thrown from a horse while racing in a rocky meadow. She suffered a cerebral hemorrhage and died on the spot.

"Do you think my father's in the best hands?"

"It really depends on the nature of his illness," I replied. "Some lymphomas are relatively straightforward, their diagnosis and treatment clear. Others can be complex and diagnosed only with sophisticated genetic and protein analysis. They require novel therapies."

"Jerry, please—no polite obliquity. It's too important. Answer my question."

We met the first week of freshman year, both having placed out of English composition and electing instead for a special seminar on James Joyce. Sarah was tall and athletic, with long chestnut hair and coltish eyes. I learned she had gone to Miss Porter's and then studied in France. Turning down Smith, where the women in her family had gone for generations, Sarah chose Barnard for the flux and grit of New York. She was initially quiet, and I wondered whether she had chosen a class over her head. Until we got to *Ulysses*.

Sarah commanded the class discussions; no one was her equal. She invoked a range of reference to history, religion, and language that cowed even the professor. She seemed to have read everything I had ever read, or heard of and hadn't read, and then authors I hadn't known about. And when challenged, she took no prisoners.

"Hugh Bisson," I began, "belongs to the older class of practitioners at the Mass General. From the social point of view he and Dr. McBride are men your father easily communicates with. And if it's an uncomplicated case, the therapy will be straightforward, and I assume McBride can handle it. Is he up to date with some of the newer treatments in lymphoma or some of the nuances of managing rare forms of the disease? I can't say. Remember, it's been more than twenty years since I was an intern and covered Bisson's patients. And I really don't know Andrew McBride."

"I hear what you're saying."

Sarah called back later in the day.

"He's on the fence," she explained. "I assume he's afraid to insult Dr. Bisson. You know the drill: they travel in the same social circles, he's our family physician, was attentive to Daddy after my mother's death. In business, Daddy is as tough as they come, but here—it's like he has no spine."

I told Sarah that her father's reluctance was not unusual. A patient fears alienating his physician and then being abandoned by him. Cowed by this fear, the patient refrains from asking for a second opinion or questioning anything.

"I'm going to ask you a big favor," Sarah continued. "Could you come up and meet with Daddy? As if you were in the area, just dropping by. It may help him to make the call to Dr. Bisson."

It was a bit evasive, the pretext transparent, but not a breach of physician etiquette. I checked my schedule and realized the next Sunday I was bringing my Toyota up to the dealer in Danvers to be serviced. Beverly was close by. Sarah could meet me at the dealership, and while the mechanic worked on the car, we'd visit with Mr. Beckwith.

The family mansion, called SeaBreeze, stood on a promontory overlooking the restless Atlantic. I remembered it from the wedding: a white security gate opened up to a long gravel road ending at the main house, surrounded by formal gardens in the English style with statuary and sculpted hedges. There were two caretaker cottages tucked discreetly at the far end of the property.

Sarah was driving a silver BMW 750i, churning up pebbles and dust with her characteristic fast pace. She waved to a pair of gardeners securing one of the taller hedges to stays in the hardening earth. It was unseasonably cool for early September. In a few weeks, the merciless winds that come off the sea would test the success of their work.

The house stood four stories and was made of blocks of gray stone hewn from local quarries. Mustard-colored shutters and pastel blue window boxes holding pansies and geraniums softened the stolid edifice.

An elderly Irish maid, dressed in a white smock and white shoes, greeted us at the wide oak door. At the wedding, I was only on the grounds, not in the house. The foyer led to a vaulted salon that was decorated with fraying Persian rugs and numerous Asian pieces: several large carved wood chests, two tripartite silk screens, blue-and-white porcelain vases that stood at least four feet tall, and a faded mural of gauzy mountains and waterfalls.

"One of my great-great-grandfathers went to Japan shortly after Admiral Perry," Sarah offered as an unprompted explanation.

I followed her up a long staircase. A series of black-and-white sketches of middle-aged men was displayed along the wall. Those at the bottom were in Puritan and colonial dress, while the portraits at the top were turn of the century. There was an obvious resemblance among them: the beefy face, bushy hair, and austere gaze.

"I called it the rogues' gallery when I was growing up," Sarah laughed. "Fourteen generations of Beckwiths going back to the Puritan settlements."

When we entered the study, Mr. Robert Beckwith sat near the far window. A sharp northern light cast a broad shadow as he stood to greet us. He was a towering man, at least six foot six, with a thick frame and fleshy neck that pressed tightly at the starched collar of his shirt. I recalled that he was a linebacker at Yale. He stood unsteadily, gripping the arm of his chair for ballast, and then strode toward us.

"Pleasure to see you again, Dr. Groopman. We did meet at the wedding, but being father of the bride it was something of a frantic affair and we didn't exchange more than a few words. Of course, I've heard much about you these many years. It's terribly kind of you to come by."

His voice was reedy, as if recovering from laryngitis, and his ice blue eyes studied me as he spoke. His complexion was sallow and I thought the globes of his eyes had a faint yellow hue.

"My pleasure to see you as well, Mr. Beckwith, although I wish it were under other circumstances."

Mr. Beckwith acknowledged my reply with a guttural rumbling, almost a growl, which I interpreted as an expression of pained agreement. He took my arm at the elbow, his commanding hand guiding me to a stiff-backed chair facing his. Sarah moved to a sofa on the side of the room, out of my field of vision.

"Something to drink?"

"Do you have club soda?"

"Mary, bring the good doctor some sparkling mineral water, please, with a lime." I studied Robert Beckwith as he turned to the maid. Below his rich groomed white hair, at the level of his right earlobe, I noted a red nodule, about the size of a dime, and a similar growth under his chin.

We sat in tense silence. I waited, feeling it best to have Mr. Beckwith direct the conversation. After the maid placed the drink and coaster on a table next to my chair, Mr. Beckwith nodded for her to retire.

"Well, I know your time is precious, so we should get down to business. It seems everyone around me has a counterproposal to my longtime friend Hugh Bisson."

Before I could reply, Mr. Beckwith picked up a thick hardcover tome and handed it to me.

"This arrived from the Merrills in Duxbury. You remember them, don't you, Sarah? Julie Merrill was at Smith with your mother."

Sarah said she remembered.

The book was the collected writings of Mary Baker Eddy.

"Complete nonsense, of course, a phenomenon of American cultural confusion," Mr. Beckwith asserted. "But the mind does take you to unexpected places in this condition."

"Many of my patients find strength in prayer. But unlike Christian Science, I think it compatible, not at odds, with modern medicine," I offered.

He looked away momentarily and then asked if my drink was to my liking. I said it was.

"Daddy, Jerry is not here to discuss theology," Sarah said as she stood from the sofa and brought a small antique chair to join us. "His input on your case is hardly in the same category as that batty Julie Merrill."

Mr. Beckwith fixed his gaze on mine, waiting for me to say something.

"I'd be glad to speak with Dr. Bisson and Dr. McBride if you'd like me to."

"Why do you think it's necessary?"

"It's always healthy to have second opinions. Due diligence. Good physicians welcome the thoughts of colleagues."

Mr. Beckwith considered my response.

"Sarah indicated you weren't certain either of my doctors is all that 'good.' "

I was embarrassed that Sarah had communicated this but tried not to show it.

"Both have practiced for a long time, and I'm sure in many instances a second opinion only confirms their approach. If the lymphoma is a common type and the therapy clear, everything could be managed locally. But if it's not, then it would be preferable to be at a major academic medical center with more specialized expertise."

"Hugh is not going to appreciate this," Mr. Beckwith said to Sarah.

"Blame me, Daddy. Tell him your headstrong daughter insisted on it."

Mr. Beckwith grunted. I wasn't sure if it signified affirmation.

"Did Sarah tell you I read American and European history, Dr. Groopman?"

She hadn't.

"Just finished Kissinger's book on diplomacy. I'm not looking for new allegiances. But let's float a trial balloon here. Give Hugh Bisson a call, and we'll see how your overture plays."

"Hugh Bisson here."

His tone was desiccated.

"This is Jerry Groopman. I'm calling about Robert Beckwith."

There was a brief silence. "Bob said you would. You know Sarah from New York."

"I do. We've actually crossed paths before, although only briefly. I was on the house staff at the General, in the seventies."

"I know who you are." His tone hardened. "I thought you were an AIDS specialist."

"I am, but I'm a hematologist-oncologist as well."

Dr. Bisson reviewed Robert Beckwith's case. Some two months ago he developed low-grade fevers after a trip to England and Scotland. His primary abnormality was an enlarged spleen, so Dr. Bisson considered an infectious cause. Like many of their generation, Bob Beckwith had had rheumatic fever as a child, and there were some residual abnormalities of his mitral valve. Infection of this heart valve can result in persistent fever and enlarged spleen. But all bacterial cultures were negative, and he was referred to Dr. McBride when Dr. Bisson suspected a malignancy.

"Andy McBride says it's an indolent form of lymphoma. We didn't even have to biopsy Bob. He had circulating lymphoma cells on his blood smear. I guess then you could call it chronic lymphocytic leukemia."

Two different terms, "indolent lymphoma" and "chronic lymphocytic leukemia," are applied to what is essentially the same disease. Lymphoma is a cancer of white cells known as lymphocytes that are found in the lymph nodes, bone marrow, and spleen. The lymphoma cells grow robustly in these niches. In some cases, they depart and circulate in the bloodstream; in this case, the disease is called chronic lymphocytic leukemia. As Dr. Bisson said, it is often indolent, growing slowly over years or even decades. It is treated with relatively nontoxic forms of chemotherapy and sometimes corticosteroids, to suppress the growth of the malignant cells and reduce the systemic symptoms associated with the disease, particularly fatigue, anemia, and low-grade fever.

"Andy McBride advised low doses of Leukeran and prednisone. As I said, Dr. Groopman, all quite straightforward."

It did sound straightforward. Mr. Beckwith's sallow complexion was likely due to his anemia and the yellow tinge to his eyes a sign of jaundice from the bilirubin pigment released from red cells destroyed in his enlarged spleen. Prednisone, a corticosteroid, would ameliorate the destructive anemia. Leukeran, a mild oral chemotherapy, would poison the lymphoma cells and reduce their numbers, allowing the marrow to resume producing normal blood.

Listening to Hugh Bisson's recitation of the case brought me back to my nights on call at Mass General. As an intern, you were paged by private doctors and given a brief synopsis of the medical history and findings and then told what would be done for the patient. Sometimes the patient did not correspond to the referring description. You learned to take nothing on faith, to verify everything on your own.

"I'd like to see the blood smear."

Silence followed.

"Bob said we simply were to talk," Dr. Bisson replied with some irritation. "You think Andy McBride can't diagnose a garden variety case like this?"

I didn't reply.

"I'll give Bob Beckwith a call," Dr. Bisson stated cryptically.

The conversation was over.

<hr>

"Some diplomat you turned out to be," Sarah said only half jokingly. "You stirred up a hornet's nest. Dr. Bisson is pissed off. He intimated that your interest in the case represents same nefarious agenda with my family."

She elaborated that the Beckwiths were major donors to Essex Hospital, tapped regularly for renovation and expansion projects.

"He said your inserting yourself in the case is the first step in the new merged Beth Israel Deaconess's campaign for money."

I assured her that was not so.

"You don't have to defend yourself to me. It's so transparent it even sounded ridiculous to Daddy."

Some three days later, Robert Beckwith sat in my office. Sarah had gone personally and retrieved his medical records and blood smear.

I first reviewed the history of his illness. There was no family history of lymphoma or other blood malignancy. I inquired about activities that might result in exposure to toxic chemicals or radiation, known precipitants of leukemia and lymphoma. His business was investments; his hobbies were golf, tennis, and fly-fishing. I asked about his gardens, since horticulturists risk contact with pesticides and other potential carcinogens. Mr. Beckwith laughed.

"I leave that to the yardmen. It's been more than two centuries since we had our own hands in the dirt."

He had traveled widely, and like his forebears, had a special interest in the Orient. There were viruses prevalent in southern Japan linked to lymphoma and leukemia of T cells and so termed "HTLV," for human T-cell leukemia/lymphoma virus.

"Andy McBride asked about toxic exposures but not viruses," Mr. Beckwith stated matter-of-factly.

I enumerated the routes of HTLV transmission: blood transfusion; mother to child during the birth process; sexual intercourse.

Robert Beckwith stole a glance at Sarah, then said he had no such exposures.

"Mother would have taken his head off if he misbehaved," Sarah added.

His physical examination proved more revealing. I palpated the red nodules on his cheek and chin. They were firm and did not blanch, and warmer than the adjacent skin.

These bumps, Mr. Beckwith stated, had appeared recently, several weeks after the onset of his fatigue and fevers. Neither Dr. Bisson nor Dr. McBride had made much of them. Both said lymphoma cells can grow as nests within the skin. Neither thought a biopsy was required, since it would only show what had been found in the blood.

I ran the tips of my fingers over the nodules again, gauging their resistance to firm palpation. Cancerous nodules are hard and resilient, while infected lesions tend to have more give. I also considered whether the nodules were a clue to the type of lymphoma. Certain kinds, including the T-cell variety related to the HTLV strains in Japan, tend to deposit in the skin. I also acknowledged that two different disorders could coexist in the nodules, one cancerous and the other infectious. Fever could come from either.

"Should they be biopsied?" Mr. Beckwith asked.

"I'm not sure yet," I replied. "I need to finish your physical exam and then review the laboratory studies and blood smear."

"The weighed words of a seasoned statesman," Mr. Beckwith said with a smile.

There were no more nodules or other findings other than his heart murmur and enlarged spleen. Curiously, there were no enlarged lymph nodes, a hallmark of chronic lymphocytic leukemia.

I excused myself and retired to the microscope room. I placed the glass slide of Mr. Beckwith's blood smear squarely on the scope's stage and began to scan it on low power. I followed a system taught in first-year medical school, initially assessing the relative proportion of white cells to reds and the heterogeneity or homogeneity of the populations. I then switched to higher magnification and step by step analyzed the morphology of the red cells, white cells, and platelets.

Blood is one of the most aesthetic tissues. Its beauty had seduced me into choosing hematology as a first career. The cells on the slide are like dancers in a grand ball, displaying costumes of brilliant colors: rich reds, aqua blues, stunning indigos. And so much can be deciphered from this microscopic world . . . the whole patient sometimes revealed in a single drop of his blood.

I confirmed the morphological changes Dr. McBride had noted in his records. Mr. Beckwith's red blood cells were small monotonous spheres instead of varied biconcave shapes. This was due to damage sustained by the red cell membrane in the diseased spleen, so the cells closed up into themselves, the way a caterpillar retracts when cut by a malicious child. The platelets, as expected, were moderately reduced in number; Mr. Beckwith was in danger of bleeding if he cut himself or fell, but not spontaneously, not without trauma. It was the lymphocytes that held my eye.

The blood smear was not perfectly prepared, the technician probably moving the drop of liquid roughly instead of in a smooth stroke across the glass. The inexpert hand had crushed many of the cells. The few intact ones I found did not look quite like classical cells of chronic leukemia. True, they had lost the variegated forms of normal lymphocytes, marked now by a monotonous visage, small and round with a condensed nucleus. But several had projections extending from their outer membrane, as if the cells were sticking out the tip of a serpentine tongue.

Was this tongue real or an artifact of the poorly prepared slide?

I returned to my office.

"Mr. Beckwith," I said, "I'd like to repeat your blood smear and perform more tests on the cells."

I explained that the slide was not prepared as I liked. I couldn't be absolutely sure about the diagnosis based on the malignant cells' appearance. I also wanted to perform genetic tests that would confirm it was chronic lymphocytic leukemia. Lymphoma, I explained, arises from the mutation of a single lymphocyte. And there were two families of lymphocytes, called B and T. These tests could determine whether the malignancy grew from a single mutated B-lymphocyte or a single mutated T-lymphocyte. The result was important, since it guided the therapy.

"I'm not sure I follow, Doctor," Mr. Beckwith replied. "Sounds awfully complicated. And frankly, sir, I don't really want to know, or need to know, T from B. I'm the kind of soldier who listens to orders and follows the captain into battle."

Many people abjure learning about their disease, believing it will lighten their burden. But, in the long run, their avoidance usually worsens the problem. Hard as it is for a patient to hear all the details, it helps to reduce the fear because the worst becomes known and the patient becomes a partner in the struggle against the illness.

So I want my patients to be fully informed. Indeed, I want them to challenge me. Their sense of their own condition is vital in directing me how to investigate their malady. Not wanting to know is often linked with not wanting to say. A headache, or a leg cramp, or indigestion, these vague symptoms in a person with cancer—a person whose immune system is weak—could signify much more than tension, or a charley horse, or a bad meal. They could indicate a serious infection or hemorrhage or growing tumor, and their early detection could spell the difference between life and death.

"Soldiers need to know what terrain they're moving into," I said.

Bob Beckwith's jaw tightened, and he drew his hands into a firm clasp.

"Frankly, I'm in a tough patch right now. Do what you have to do. Tell me where we're headed, and spare me the details."

⌒

The second blood smear had a broad "feathered edge," and the cells were not distorted. Many now showed fine projections, resembling not serpentine tongues but unkempt wispy hairs. It was this image that gave a rare blood disease its name: hairy cell leukemia.

There are an estimated six hundred new cases a year of hairy cell leukemia in the entire United States. Many hematologists and oncologists therefore may see only one or two of these patients over the course of a career. I had cared for more than a score during my fellowship training in hematology and oncology at UCLA. The director of the program, Dr. David Golde, was keenly interested in this rare malady, and patients were referred from all over the country. After my clinical years, I entered David's laboratory and studied a variant form of hairy cell leukemia involving T-lymphocytes.

I took the slides for confirmation to Brad Sherbourne, a pathologist specializing in blood disorders. He confirmed my initial impression and then stained one slide for a special enzyme called TRAP. The malignant lymphocytes were found to contain it. This histochemically verified that the diagnosis was hairy cell, rather than chronic lymphocytic leukemia. The protein and genetic tests returned later in the week showed it was of a B-lymphocyte type. I had suspected T-cell, possibly due to HTLV, but was wrong.

Robert Beckwith's hairy cell leukemia could progress slowly or explode into an aggressive form with fatal complications. One of these complications was opportunistic infection, since immune deficiency is a hallmark of hairy cell leukemia. I wondered again whether the red nodules represented, as Dr. McBride assumed, deposits of malignant cells or an opportunistic infestation.

There was much new information, and my immediate impulse was to call Robert Beckwith.

But I decided to contact Dr. McBride first. This adhered to the imperative of diplomacy. Dr. McBride was the specialist who had arrived at the initial diagnosis. He also might be more successful at communicating the "details" Mr. Beckwith did not want to hear.

"Interesting," Dr. McBride muttered. "Of course, my treatment plan of Leukeran and prednisone covers that base as well."

Leukeran, the mild oral chemotherapy drug for indolent chronic lymphocytic leukemia, was also once used in hairy cell leukemia.

But it was now regarded as a third-rate drug. And prednisone, a powerful corticosteroid, was potentially dangerous.

"Prednisone is controversial in hairy cell, as you know," I stated, keeping my tone friendly. "It can further reduce immune defenses and bring on opportunistic infections."

"But his spleen is enlarged," he replied, his tone sharpening, "and chewing up his red cells. He's quite anemic."

"I'd hold off on prednisone for now. The anemia can be managed with erythropoietin, if his serum levels are low, or with transfusions. It should ultimately reverse with effective treatment against the underlying leukemia as the spleen shrinks back to normal. And instead of Leukeran, I would consider a more definitive, newer agent, like CDA."

A heavy silence hung between us.

"Transfusions are messy and inconvenient, and erythropoietin—I haven't used it to boost red cells. Don't think much of CDA. I know you younger guys like to run after every new development that comes along, but there's a lot to be said for experience." He paused, and in an ethereal tone said, "I remember my last case of hairy cell two years ago—an Italian fisherman from Gloucester, nice fellow, good wife, pack of kids. His brother—who shared the same boat—his brother insisted on going down to Boston to the Mass General. They gave him CDA. Wiped out his immune system. An infection killed him."

"CDA has the side effect of immune suppression," I agreed, "but that can be ameliorated with the new blood cell growth factor G-CSF that boosts the white count. And, as you know, for the majority of patients CDA is *curative* therapy. Leukeran is *palliative*."

McBride seemed to be listening, so I continued, although I feared sounding like a lecturer.

"Hairy cell leukemia itself, of course, causes immune deficiency. The leukemic cells are unable to defend against infection, and they crowd the marrow and block production of healthy functioning white cells. Your fisherman's demise may have been due to the underlying hairy cell leukemia, not the CDA."

"I do, of course, appreciate your thoughts. But I'm not going to debate you, Dr. Groopman," Dr. McBride answered tiredly. "Leukeran is the best option here." Almost as an afterthought, he said, "I'll hold off on the prednisone."

I had at least won half a victory. Perhaps I should have been satisfied with that. But I couldn't disengage, so I tried again, this time imitating his style.

"I'd appreciate your thoughts on those red nodules, Dr. McBride. I'm not sure what they are. They remind me of one of my cases, at UCLA in 1979, when I was a fellow. We had a special clinic devoted to hairy cell leukemia. An elderly farmhand who worked in the San Joaquin Valley was evaluated with low-grade fevers and skin nodules. I assumed they were chloromas as well and didn't send the man to a dermatologist for a biopsy."

Chloroma is an old term for solid nests of leukemic cells within the skin, the assumption that Dr. McBride had made about Mr. Beckwith's red nodules.

"So we didn't evaluate them further. Of course, CDA did not exist, and we gave Leukeran. After a few weeks on chemotherapy, the patient had a raging fever. The nodules harbored an unusual fungus. We intervened with antifungal drugs, but it was too late. He died of the infection."

I felt uncomfortable confessing one of the most painful experiences I had had to Andrew McBride. It still haunted me. Although I accepted that a doctor cannot pass through a career without making mistakes, I had never forgiven myself. I believed, perhaps, that forgiving would mean forgetting and make me prone to more unwarranted assumptions.

Colleagues reassured me that the patient "would have died anyway." The farmhand was infected with a fungus that is often fatal, even with early treatment. I listened and pretended to be consoled but inside refused to rationalize my lethal error.

"I'll keep your California experience in mind and try to do what is best for Bob," Dr. McBride said politely.

⌒

"Now you've got Dr. McBride riled," Sarah reported.

I was surprised, I replied, since we collegially discussed the diagnosis as hairy cell rather than run-of-the-mill chronic lymphocytic leukemia, and I very diplomatically raised the possibility of an infectious cause for the fevers and red nodules.

"Well, he's done a good job of frightening Daddy. He said you wanted to use some horrible toxic chemotherapy. And then he confided that you had made some terrible mistake with a past hairy cell leukemia patient and the patient died because of it."

I was furious and for a moment speechless.

Why did I assume McBride would accept me as a colleague rather than the threat I was? His behavior, in some way, was made all the more loathsome because of the façade he assumed: the upper-class gentleman, refined of speech, expressing the desire to do what's best for his patient's welfare.

It was no longer a simple skirmish. It was war.

For a moment I considered withdrawing. Sarah was a close friend, but Robert Beckwith was a person with whom I had no close link or affinity. Who needed this aggravation? I could foresee more time and energy spent dealing with Bisson and McBride, not to mention coaxing Mr. Beckwith through a complicated and uncertain leukemia. I had many other patients to attend to, much work waiting in the laboratory—I could score points in heaven without martyring myself on their backstabbing knives. But I knew I could not walk away from a patient in need. It contradicted everything that made me a doctor. And, I admitted to myself, I was not a person to shy away from confrontation. As a kid, I was tall and hefty, and growing up on the playgrounds of New York City taught me how to hold my own. Some fights were settled with fists, others with cutting words, none by appeasement or running away.

What to do? McBride's treatment plan was not malpractice per se. He had dropped the clearly dangerous drug, prednisone. Leukeran was simply third-rate. My approach with CDA could be curative but, it was true, riskier. And if it failed, I was sure to be blamed by Drs. Bisson and McBride and, I worried, by Sarah as well.

Failure. Isn't that always the burden a doctor labors under? Failure not because of error but because of the nature of disease. The reality that doing your best, covering every base, responding to every evolving aspect, is just not enough. Was my worry the soft whisper of ego—that as a physician you want to look good, sustain a reputation as a winner, score as many victories as possible?

"Jerry, my father's life is on the line," Sarah forcefully stated, interrupting my reverie. "I don't give a damn about Bisson and McBride. I told you I want him in the best hands, getting the best treatment."

I replied there were many expert hematologists and oncologists in Boston who would fulfill that imperative. I was tainted in her father's mind and could recommend a specialist at the Mass General, where her father might be more comfortable.

Sarah paused briefly.

"No. It would be a third and unknown cook in an already messy kitchen. I know you and believe you're right for Daddy. He just can't think clearly now."

"We do need the third opinion," I asserted. "From a respected specialist and not one who is my close friend or is in any way prejudiced toward my advice."

"What if Bisson refuses? He was pissed off from the start, when I asked about what I read on the Internet."

I explained then her father would be forced into a difficult position—having to choose without an independent arbiter.

We talked further. Sarah confided she had again searched the Internet, under "hairy cell leukemia," when I made the diagnosis. She had collated a list of papers on the disease. I listened to her recite the authors and told her who were my direct collaborators or friends.

A week later Sarah accompanied her father to the Mayo Clinic in Rochester, Minnesota. The hematologist there was an established clinician who had reported on that institution's experience with the disease.

I received a call from him the day after their appointment.

"I've already spoken with his primary physicians, Dr. Bisson and Dr. McBride. There are always pros and cons to choosing a riskier but more definitive therapy like CDA over Leukeran. I come down on the side of CDA. Mr. Beckwith is still a vital man, active, a young seventy-three, with no other serious medical issues. He should tolerate the therapy and its aftermath."

The Mayo specialist said a detailed note would follow, addressed to Dr. Bisson, with a copy to me.

I thanked him and called Sarah.

"I just don't understand him," she said. "What more does Daddy need to make the right choice?"

He was still wavering, the specter of severe immune deficiency and fatal infection looming in his mind. It wasn't like him to be paralyzed, Sarah said. If anything, he was a person who sometimes decided too quickly and with too much confidence.

"We need someone else to convince Daddy, someone whom he would respect. He won't listen just to me."

I thought for a long while, and then a possibility leapt into my mind.

"There's a patient I consulted on who received CDA—not for

hairy cell but for another type of leukemia. He's a savvy and well-known person, a person of integrity."

I explained that I needed her father's permission to disclose his illness to this third party; otherwise, it was a breach of confidentiality.

"I'll get it."

Three days later my call to Bill Martin was returned. Bill had been traveling in Asia—exactly where and why, he did not say. Now in his late seventies, he had been in the OSS—the precursor to the modern CIA, during World War II, and then an undersecretary of state in a later Republican administration. Still a counselor of high officials, his name occasionally appeared in the newspapers as one of the last so-called wise men. Bill had come to our hospital from Connecticut to receive CDA when it was still an experimental therapy and not available locally. It was now some three years since he was without evidence of disease—too early to declare he was cured, since a follow-up of at least five years was required. We were nonetheless hopeful.

"I understand," he said in his measured voice, prolonging the last syllable of the verb, the *a* broad, almost British in inflection. "My pleasure to speak with Mr. Beckwith."

It was more than a trick or clever attempt to trump Drs. Bisson and McBride with a person of greater stature than they could summon from the North Shore set. One of the most beneficial conversations for frightened and despairing patients is not with their physician but with a person who has gone through what they are facing and can speak in a way no one else can.

Bill Martin informed me later that day the conversation had been "extensive, detailed, frank, and highly productive."

I hadn't realized Bill had graduated from Yale some five years before Bob Beckwith.

"I told him for a fellow Yale man it was a 'no-brainer' who should navigate his ship through rough seas to a truly safe harbor. And not to worry about his local doctors. The same reasons they have to keep him there now they will have to take him back in the future."

⌒

"Who are all these people?" Bob Beckwith growled.

"They're my team for the month," I explained, introducing the oncology fellow, Carol Goldstein, the resident, Mark Murphy, the interns, John Slater and Yolanda Drew, and two Harvard Medical School students, Valerie Nguyen and Kwame Brooks.

"Do I have to be on display for the United Nations?"

"Father, please!" Sarah exclaimed. "Jerry told you this is a teaching hospital, and all his patients are cared for by him with residents and students."

I assured Mr. Beckwith we would limit the inconvenience. He would not have to repeat his history more than one more time. But I also explained he would be examined by all of the team members. This might seem annoying and excessive, but there was a real advantage. Sometimes a physical finding previously missed was picked up by one of the house staff or students. And if problems occurred during the middle of the night, there would be one or more doctors on call who knew him and his case well. This multilayered care provided safety nets, despite the sense of being pawed.

"I'm not keen on mistakes, Dr. Groopman."

"Please call me Jerry."

"Jerry it is. I still am very, very anxious about mistakes being made."

I decided it was time to assert myself and to instill confidence. I would do this honestly, and without, I hoped, appearing defensive or vindictive about Andrew McBride.

"Bob, you know no one is perfect . . ."

"But I want you to be. At least, in my case. No errors."

"I'm sorry, but no one can promise that. I'll do my very best, as I do for everyone I care for."

I paused, searching Bob Beckwith's face for a response. There was none.

"I have considerable experience with your disease. It is relatively rare, and there's a lot to be said for having seen a score of people with a rare illness, rather than a handful. There is clinical intuition that develops from managing so many patients. This intuition helps protect against mistakes."

I wasn't certain if what followed would solidify our relationship or act as a wedge.

"The correct diagnosis of your condition was not made in Essex, as you know. I was not satisfied with the quality of a very simple but important test, the blood smear. If you had been treated as originally intended, it would not have been optimal. In fact, prednisone, a corticosteroid, is problematic in hairy cell leukemia. It can often worsen the immune deficiency and accelerate severe infections."

Sarah sat pensively at the side of the bed, her eyes trained on her father's face. I noted her gaze tighten and her brow drop.

"I was the one who prompted Dr. McBride to biopsy the nodules on your skin. That biopsy revealed not only the hairy cell leukemia invading the dermis but also a smoldering infection with *Mycobacterium kansasi*. That's a very serious bug in someone with your condition, related to tuberculosis. It could have spread to your liver, lungs, and brain if you hadn't been treated with powerful antibiotics.

"I insisted on the skin biopsy because of the mistake I made in the past, with the farmworker in California. I try to learn from my mistakes—the worst way to learn. I'm not aware of many significant mistakes in my career, but the ones I've made I can describe chapter and verse."

I paused to focus my thoughts, wondering if I was going on too long.

"I'll do my very best to pay attention to every detail, to care for you without making a mistake. But I repeat: I cannot give you a guarantee. And the team that stands before you—I check on them, and believe me, they check on me, even though I'm the boss. That's the way it works in a teaching hospital. These young doctors and students are not passive yes-men and -women. They are taught to challenge their seniors, to criticize constructively, to think on their own. That doesn't threaten or annoy me—like some physicians you may know who are reluctant to consider other opinions."

I had said enough. Probably too much. But I was passionate about the care of patients. I was not an ivory tower snob who dismissed doctors and hospitals in the community, outside the academic medical centers, as prima facie inferior. In fact, I had a special admiration for those who served their patients expertly in the community without the extra pairs of hands and eyes afforded by the residents, students, and senior specialists I called on to help me.

"Makes sense to have a platoon of juniors around," Bob Beckwith allowed, "so long as I know you're the general in charge."

I assured Bob Beckwith I was and one with a tight grip on the reins.

The first day we began a prolonged infusion of CDA. The chemotherapy nurse reviewed in great detail with Mr. Beckwith and Sarah that drug and all the other medicines he was receiving, their potential side effects, and the importance of reporting any and all symptoms he might have to the nurses and the doctors. I later heard from the nurse that Sarah had listened attentively, jotting copious notes in a diary. But Mr. Beckwith found it hard to concentrate and told the nurse her words only increased his anxiety.

Later that day, I returned to his room, again with the residents and students in tow. I sat down next to his bed, the medical team standing behind me in a semicircle. The late-day autumn sun had warmed the green vinyl hospital-issue chair and made it pliant.

"I understand how frightening it is to hear about potential side effects of CDA and then today to begin the actual treatment," I began. "But it's important to be the boy who cries wolf each time you feel something different or unusual." I locked my gaze firmly onto his. "Unless you're aware of the possible side effects, you may not report them when they first occur. And it will be harder for us to intervene and help."

I paused. Sarah's visage was heavy with worry; Mr. Beckwith's had a blank, noncommittal look.

I assessed him as a stoical person, one reared in a tradition where emotions are muted and complaining about one's pain is judged bad form.

"Bob, something is likely to happen—an infection or a hemorrhage particularly. It's the reality of the disease and its treatment. It's critical that you communicate with us. We cannot read your mind. Your complaints will direct us where to look and how to remedy a problem. You're part of the team we hope will effect a cure."

He said nothing in response, only vaguely nodding in what I hoped was assent.

Bob Beckwith was discharged from the hospital and given an appointment to return to the oncology clinic in two weeks. G-CSF, to boost his white cells and lessen the immune deficiency, and erythropoietin, to boost his red blood cells and reduce the need for transfusions, were to be given by a private visiting nurse. Bob didn't feel confident he could inject himself with the drugs, although most patients do, the needles and syringes being similar in size to those used for insulin.

The nurse reported to me later that week that the low-grade fevers continued and that Mr. Beckwith seemed lethargic. The infected nodules on his skin had shrunk significantly. He was tolerating the antibiotics against the *Mycobacterium kansasi*.

I called Dr. Bisson, whose office was drawing Mr. Beckwith's blood tests during the interval before the next visit in Boston.

"He's been through the mill," Dr. Bisson reported. "Poor devil.

First Claire's accident, now this. I'm going to prescribe an anti-depressant. I like Prozac. Andy McBride felt nothing about the leukemia or your chemotherapy would contraindicate it."

I thought for a moment, double-checking in my mind whether there were any significant drug-drug interactions between Prozac and the antibiotics. I couldn't think of any but was wary of relying on memory.

"Let me check in the *PDR*."

The *PDR* stands for *Physicians' Desk Reference*, a compendium of some two thousand pages detailing available drugs, their indications, doses, side effects, and metabolism. I reached for the heavy volume and read the relevant section.

"Seems safe." I averred that just because an adverse interaction wasn't noted in the *PDR* did not mean it couldn't occur. It only signified that one hadn't been reported to date.

I asked Dr. Bisson to keep me informed of Mr. Beckwith's response to the Prozac, and he said he would.

The follow-up appointment in our oncology clinic two weeks later was on an atypically warm November morning. It was past New England's Indian summer, and yet an embracing sun held the city in a pleasant grasp. Bob Beckwith was dressed informally, in khakis and a white button-down oxford shirt, with a white boating cap covering his balding head. He was already losing his hair from the chemotherapy.

"Daddy, you're still feeling pretty low, aren't you?" Sarah began.

I watched Bob Beckwith's response closely. He looked tired and drawn. His normally sharp gaze was absent. He was slow to reply but finally did.

"Still in a tough patch."

He almost slurred his words.

"Are you taking the prescribed dose of Prozac?" I asked, signaling to Sarah that I was concerned he wasn't. Medications like Prozac can cause drowsiness and incoordination if taken in excess.

Again, the reply was slow in coming.

"Think so. My lovely Irish girl checks on all my medicines."

"You mean the visiting nurse?" I wanted to be sure he didn't mean Mary, his maid.

Bob Beckwith nodded.

"Are you dizzy or having difficulty in thinking clearly?"

"I've had trouble thinking clearly all my life."

Bob Beckwith smiled at his own joke, a wide, almost silly grin. I smiled back.

I worried that the drug was building to excess levels in his blood because of some unreported interaction with his multiple antibiotics.

"Daddy hasn't been himself, since he learned from Dr. Bisson he had some type of cancer," Sarah said. "He's weak from the anemia. He tripped on the dock the day before we came to Boston for our first meeting with you. Now he cuts himself shaving every morning because his hand shakes so much. But he refuses to let Mary shave him."

Sarah smiled admiration for her stubborn father. He looked blankly back at her. I glanced at his neck and noticed a few small scabs where he was nicked by the razor.

"Does he bleed for a long time from the cuts?"

Mr. Beckwith's platelet count was low from the leukemia and the chemotherapy. It would take some weeks to recover if the treatment was successful. Before then he was at risk for hemorrhage.

"They do bleed, but Daddy has some styptic pencil, you know, that old-time astringent."

I told Bob Beckwith my father used the same remedy when he cut himself shaving.

"But I laid down the law this week," Sarah continued. "He's too unsteady, bumping into things around the house, to go out unaccompanied. And I told Mary no guff from him—she should shave him. In answer to your question, it seems the Prozac has worsened things."

I made a note in his chart of these events and glanced at the clock. We had already passed twenty minutes, the usual time for a routine follow-up exam. There were three other patients on my clinic roster still to be seen. His anemia, general weakness after the chemotherapy, and depression, as well as side effects from the Prozac, could all account for his difficulties. I would speak later with Dr. Bisson and suggest Mr. Beckwith be seen by a local psychiatrist to monitor the medicine, or perhaps switch to a different antidepressant.

"I'm going to exam you," I said to Mr. Beckwith. "Should I ask Sarah to leave?"

He said it wasn't necessary.

Sarah stepped forward to help him with his clothes, but he waved her off. He slowly undid his khaki pants and then the starched oxford shirt. He struggled with the buttons but finally was down to his boxer shorts. I draped a blue cotton hospital gown over him. He sat on the examining table, sad and defeated, his broad shoulders sagging and his trunklike limbs hanging limply at his sides.

"Reading anything interesting?"

I was trying to engage him in his stated passion for history.

"It's been hard to read."

"It is when you're ill. Jaunty hat," I joked. He was still wearing his white boating cap. "I could use one to cover my pate. But your hair will grow back after the chemotherapy. I'm stuck."

Sarah explained her father particularly disliked losing his hair, a part of his appearance he'd always prized, and wore the hat all the time. Even in the house.

"Vanity is a healthy sign when you're ill," I said, pleased that he retained interest in his self-image despite the depression.

I examined him systematically, moving down from ears, eyes, and mouth to neck, chest, abdomen, and limbs. There were a few small bruises on his shoulders and shins, probably from the fall at the dock. Neurological assessment showed his reflexes were brisk and his strength intact. There were no abnormalities in sensation. His coordination was slightly off when I asked him to walk heel to toe, like a tightrope walker, but many ill people find this difficult to do without some listing. The so-called cranial nerves, which control the movements of the eyes, cheeks, and tongue, were normal.

The last parts of his anatomy to inspect were his genitalia and rectum. His boxer shorts were still on.

I had Sarah leave the room.

Mr. Beckwith struggled to remove his underwear. I offered to help, and he did not object, merely wincing at the indignity. There were a few more bruises on his buttocks, which appeared after the fall, but no other findings.

I helped remove the hospital gown and held his shirt for him.

There was a knock on the door, and Nancy Rao, the chemotherapy nurse who had given Mr. Beckwith his CDA treatment, entered the room. Sarah followed.

"Good morning, Mr. Beckwith," Nancy greeted.

Bob Beckwith's visage brightened and he gallantly took off his hat. "Good day to you, young lady."

The three began to converse about the sparse catch this season. Nancy's father was a Portuguese fisherman who worked out of Provincetown.

I jotted down my note in Mr. Beckwith's chart, underscoring the question of side effects from the Prozac, and then turned to thank Nancy for dropping in. As I did, my eyes were arrested by a blue-gray bump on Bob Beckwith's cranium.

I quickly stood and inspected the area. The conversation stopped.

"What's wrong?" Sarah asked.

I did not immediately reply. I ran my fingers over the fleshy mound. It was two inches in diameter, not warm but easily compressed.

"When did this appear?" I asked. It had not been present on his initial examination.

Mr. Beckwith didn't know. That would be some two weeks earlier.

My stomach tightened and pulse quickened. I silently, fiercely berated myself for examining Mr. Beckwith with his hat on. I had thought I was doing him a good turn, supporting his vanity. I had broken discipline.

With his low platelet number, even minor injury to the head could cause internal bleeding under the skull. The blood would collect under the fibrous sheath called the dura that encases the brain. This was termed a "subdural hematoma." It was insidious in its presentation: sometimes the patient only had a vague headache, or showed a subtle change in personality. Like many things in medicine, you had to think of it to make the diagnosis; once you did, it seemed painfully obvious in retrospect.

A subdural hematoma, if not drained, could cause mounting pressure until the brain collapsed onto its supporting stalk. This event, called a cerebral herniation, cuts off vital neural circuits that control breathing and heartbeat. The patient dies within minutes of the collapse.

"I'm going to get a CAT scan of your head," I said to Mr. Beckwith.

Sarah's eyes widened. Mr. Beckwith looked mildly perplexed.

I rapidly stated my suspicion, that the fall at the dock, in light of his low platelet count and tendency to bleed, caused a subdural hematoma.

"It's like having a frontal lobotomy, in terms of what it can do to your personality," I explained.

"And you think it's been there all this time, two weeks, and no one, not Dr. Bisson who saw him in Essex last week, or you, considered it?"

I faced the question head-on. "Yes. I think the hematoma has been developing since his fall. And I almost missed it."

I was tempted to elaborate that often it takes time for such personality changes to be manifest, and occurring against a background of severe illness and Prozac, it was easy to overlook. But my reply sounded defensive as it aired in my mind. I had to admit that the discipline I taught my students and interns, to examine everything and to create a list of diagnoses, beginning with the common and obvious and extending to the rare and subtle, was meant to protect against just this sort of situation. Doctoring was a balance between the sixth sense of intuition and the tedious reiteration of diagnostic lists. You didn't have to be brilliant to be a competent doctor, but you did have to be thorough.

I sat in the darkened anteroom watching the CAT scan machine. The huge metal apparatus whirred around Mr. Beckwith's head. His skull and brain appeared in crosscuts on the computer screen before me, the inner tissues revealed as if they had been filleted. A large dark mass with the density of blood extended over the surface of both frontal lobes and was deeply indenting the underlying brain tissue.

I paged Sven Norquist, one of the hospital's neurosurgeons. He was in the operating room. I was connected to him by a speakerphone and explained the case. He said that as soon as his current surgery was wrapped up, probably in less than an hour, he'd review the CAT scan himself and then find the Beckwiths to explain in detail what needed to be done. I then contacted the hospital admitting office and arranged for a room.

I sat for a few minutes after the CAT scan was completed. I was still furious with myself. I had become distracted, for just a moment, and it could have resulted in disaster. I had made an assumption, that Mr. Beckwith was depressed and possibly overmedicated, and I failed to think beyond that. If Nancy Rao had not come into the room, and Mr. Beckwith had not been bred to take off his hat in front of a lady, or if that custom had been lost along with other cognitive responses because of the blood over his frontal lobes . . .

Enough, I instructed myself. Replaying the incident over and over served no purpose. I had to put it to the side, still present in my mind but not at the forefront, or it would dominate my thinking and paradoxically impair my functioning.

Some three hours later, I stood to the side of Sven Norquist. I wouldn't ordinarily go to the OR but my near error still weighed heavily on me and I was anxious to see it safely resolved. Sven, using a high-speed drill, made two holes in Bob Beckwith's cranium. Fresh blood poured from the opened skull, it being a bone filled with small fragile vessels. Sven applied a thick paste of a material called bone wax to stanch the flow. He then cauterized the sheath of dura with an electric probe and, satisfied it was coagulated, made his incision. A viscous liquid the color of crankcase oil spurted out. He placed a low-suction drain into the hole, careful to temper the outflow and very gradually relieve the pressure on the brain. If the brain reexpanded too quickly, it could swell and trigger a stroke. After less than a half hour, Sven was satisfied that the subdural hematoma had been safely evacuated. A tube was left to drain under gentle suction for twenty-four to forty-eight hours to remove any residual subdural fluid.

Later that evening, I visited Mr. Beckwith in his room. The lights were dimmed. The green spike and curve of his EKG tracing raced on the monitor over his bed. He was sleeping, his shaved head bandaged, the tube holding a small amount of the yellow-brown subdural blood. Sarah was seated at his bedside, a sheaf of financial spreadsheets in her hands.

"Hard to really concentrate on this," she said as she put the papers aside.

I nodded.

"You know, intellectually, that someday you'll take care of your parents. But when it happens, it's still strange."

I pulled up a chair and put my arm around her shoulder.

"You know what worries me the most? That I will have to make decisions for Daddy. That one day you will come to me and ask whether the plug should be pulled."

Her rich blue eyes became moist with tears. I tightened my grasp.

"If we reach that point, and I hope we don't, decisions like that are never made alone. I'll be there for your dad. And for you."

Sarah smiled wanly in acknowledgment of my words.

"Try to keep your mind from reaching too far ahead, Sarah. The cliché about taking things step by step is true. It's hard, I know, especially for people like us who are used to planning and strategy, trawling over every scenario. But in this instance, it's best to keep focused on the present and not pursue every form of the future."

Sarah paused gravely.

"If that blood on his brain hadn't been found, there wouldn't have been a future."

The drainage of the subdural hematoma was uncomplicated, and Bob Beckwith regained his former personality. We stopped the Prozac. About a month after the surgery, he developed a high fever and cough. Sarah immediately brought him to the hospital, and an incipient pneumonia was diagnosed. It proved to be from *Haemophilus influenzae*, a bacterium that often causes disease in hosts with impaired B-lymphocytes. The infection was extensive, and for a while I feared he might need a respirator. But by the fifth day of intravenous antibiotics he rallied. After three weeks, he returned home on oral therapy.

Hugh Bisson made regular house calls and called me to report after each one. Our conversations were polite and professional, and I received copies of all lab work and notes from his office without having to request them.

"What are your plans for Christmas?" I asked Bob Beckwith on a follow-up appointment.

"Very quiet, Jerry. Sarah will be over. I'm still exhausted from the siege with that pneumonia. And your family? I remember how excited Sarah used to be trying to guess what would be in her stocking. Your children must be in a state of high expectation, especially your little girl."

I explained that the kids were given gifts on Hannukah, which had been some two weeks earlier.

"Well, the gift you've given me, Jerry . . ." Mr. Beckwith's voice wavered. "I don't know how I can repay you."

"It's not a question of repayment."

When I looked at his blood counts two weeks ago, a delicious joy began to suffuse my core. Robert Beckwith's life likely had been saved. His last blood smear and marrow showed no evidence of residual hairy cells. I knew I had to temper this feeling. It would take a year for his immune system to recover and several more years to be certain that the leukemia had been eradicated. There was also the risk of a second cancer; people with hairy cell leukemia had about a 5 percent chance of this, and chemotherapy somewhat enhanced its incidence. Still, more than three out of four patients with hairy cell leukemia were cured by CDA, and Bob Beckwith had a good chance to be one of them.

"Would you like to join Sarah and me with your family on Christmas Day? Great sledding for the kids on the local golf course."

I thanked him warmly but had to decline, explaining that my wife and I usually made rounds for Christian colleagues on the holiday. Perhaps another time.

⌒◠

Christmas Eve began bright and chilly. I had only a short glimpse of the morning, since I covered the AIDS service. My wife, Pam, left early as well. She was on call for her associate. I returned home before her, as the late-afternoon light diffused into a soft yellow.

"A large package came," my second son, Michael, informed me as soon as I stepped in the doorway.

On the floor of the foyer was an opened brown box. Bubbled wrap was strewn around it, along with shreds of brightly colored wrapping paper.

The gifts for the children were clearly chosen with Sarah's insider knowledge. Our older boy, Steve, a techie, had received a computer strategy game that, he explained, was just released. "I already loaded it. *PC Gamer* reviewed it this month. It's very cool. You pose as an adviser to Stalin, but you're really a British spy." Mike had been given a set of calligraphy pens and an accompanying book on Chinese characters. Emily was already wearing her flowered-print jumper. It was a little long in the sleeves, but it would be perfect by spring. I left Pam's gift for her to open and picked up mine. I could tell it was a book even before carefully removing the wrapping. The card was Bob Beckwith's personal stationery, Tiffany cream-colored paper embossed with initials in royal blue, written in his elegant script: "In deep appreciation for helping this ancient mariner navigate such difficult shoals." The provenance stated it was an early edition of Joyce's masterpiece.

Grandpa Max

Columbia's Neurological Institute stood grandly against an azure sky on a September morning in 1980. I held my grandpa Max's hand as we moved in the bustling current of nurses, interns, staff doctors, patients, families, and students down 168th Street. We ascended the ramp to the institute's entrance. The elegant brass letters over the marble archway, the subdued tiled floors of the lobby, the troika of elevators were all known to me and comforting. I had been a medical student at Columbia. Several of my former teachers had alerted Dr. Edward Mathers about my grandfather. The buoyant sense of being special, and coming to a renowned institution, lifted my hopes.

The past weekend, my mom and stepfather, Maurice Pollet, had gone to see my grandfather at his house in Rockland County. They were summoned by my great-uncle Sam. After the death of my grandmother Rose in 1976, my grandfather had changed. Over the first year, he became increasingly morose and distant. Most of each day was spent in his room listening to favorite operas. He went out only occasionally, to attend synagogue or a special event at the seniors' club. We attributed his behavior to the loss of a beloved companion after fifty-three years and tried to engage him in family outings. But his mood did not lift.

Then we discovered he was giving away valuable possessions.

Grandpa had always had a large and generous heart. His acts of charity were legendary. But when queried about where the silver wine goblet used for the blessing over the wine on Sabbath dinners went, or who received an antique pocket watch that had been handed down for two generations, he responded with a blank expression.

Recently, his actions had taken a bizarre turn. He ripped his bedsheets and left the debris on the floor. A large down quilt was picked apart, its feathers strewn in the corner of the den. Several mornings he forgot to remove his pajamas before putting on his shirt and pants.

A few days before, Uncle Sam, a bookish bachelor who had lived with my grandparents for decades, entered Grandpa's room to clean up yet another shredded bedsheet. Grandpa had brandished a knife.

"Uncle Sam is terrified," my mother told me in a distraught voice.

Grandpa Max had loomed larger than life when I was growing up. He ventured, in fantasy and fact, outside the usual world of immigrant and first-generation Jews. He recounted lurid tales of taking San Juan Hill with Teddy Roosevelt and the Rough Riders, and marching triumphantly down the Champs-Elysées with General Pershing. His poor eyesight, I later learned, made him 4-F when he tried to enlist. He claimed to help edit his favorite author, Robert Louis Stevenson, and recited by heart long passages from *Treasure Island, Kidnapped,* and *Dr. Jekyll and Mr. Hyde.* On my eighth birthday, he called, and in a gruff stage voice, announced he was the keeper of the Bronx Zoo. A buffalo was being shipped as a special present for me, Max Sherman's special grandson. Was my basement large enough?

Grandpa was a strapping figure, six foot two, with broad shoulders and viselike hands. "That comes from hauling mailbags," he told me. During his youth, he worked in the post office. He never lost those powerful arms, even after leaving the mail job to run a pharmacy. He kept limber by fishing off Sheepshead Bay or Montauk Point with his many buddies. These fishing friends were construction workers, plumbers, subway motormen. On Sunday, they became "pirates," their treasure being cartons of beer and satchels of sandwiches. Despite his strength and taste for physical exertion, my grandfather had never been violent or threatening in his eighty-four years.

Uncle Sam conformed to the family norm of his and my grandparents' generation. His passion was books, and his room was filled with tomes in Hebrew and Yiddish. He had helped my grandmother Rose care for their mother, my great-grandmother Miriam, when she became senile. Grandma Rose and Uncle Sam could not countenance a nursing home. And years later, as Grandpa became more disorganized and disheveled after Grandma Rose's death, Uncle Sam devoted himself to his care. But that could no longer continue.

My mother informed Grandpa she was taking him on "a trip" to Manhattan. He didn't ask further. I flew down from Boston to attend the consultation with my mother and stepfather.

Dr. Edward Mathers was considered New York's premier neurologist in the diagnosis and treatment of dementia. He was tall and handsome, in his late forties, with thick black hair and aristocratic features. In the breast pocket of his starched white coat was a Montblanc pen.

"Pleasure to meet with you all," he said.

Dr. Mathers shook my hand and then my mother's and Maurice's.

The clinic consultation room was small, its walls white and undecorated. It held several wooden chairs, a compact chrome-and-Formica desk, and a stool next to the examining table. A single window looked west to New Jersey.

Dr. Mathers sat at the desk, opened a thin folder, and uncapped his pen. My grandfather took a chair next to my mother and Maurice. I pulled the low stool from the corner and sat between Grandpa and the neurologist.

"What is your name?" Dr. Mathers asked Grandpa.

"Max."

"Max what?"

My grandfather did not respond. He slowly moved his gaze from Dr. Mathers to the window. I followed his line of vision. A large tug-boat churned its way down the Hudson River.

"Max what?" Dr. Mathers asked pointedly.

"Sherman," my mother said.

"He answers, not you," Dr. Mathers instructed.

"Now, Mr. Sherman, look at me. Who was the first president?"

"George Washington," Grandpa quickly replied.

"Who is sitting in the room with you?"

"You are."

I stifled a nervous laugh.

"Who else?"

"My daughter and a grandson."

"Name them."

"Muriel and her boy."

"What's her boy's name?"

Grandpa hesitated, then looked hard at me.

"Jerry."

"And what is your wife's name?"

Grandpa's eyes drifted again to the window. The tug had passed.

"Your wife's name?" Dr. Mathers repeated insistently.

"It was Rose. She died three years ago," I said gently.

A cold pain gripped my heart.

Dr. Mathers studied Grandpa's face for a long moment and then began to write in the chart. As he wrote, not lifting his head, Dr. Mathers asked my grandfather to count backward from one hundred.

"A hundred, ninety-nine, ninety-eight . . ."

Still writing, Dr. Mathers muttered, "Continue."

But my grandfather seemed uninterested after ninety-four. He had found a loose thread on his knitted gray vest. He began to pick at it, and as the thread extended from the fabric, he abruptly stopped.

"Rose will scold me for this." He smiled mischievously. For the first time that day there was light in his gray eyes.

Dr. Mathers ignored the remark and told my grandfather to get undressed to be examined.

My mother helped take off Grandpa's clothes. His exposed body looked pale and doughy. When I was a little boy and went on fishing trips with him, my parents would deposit me at his Bronx home the evening before so we could leave before dawn and arrive at Sheepshead Bay for the first boat. Although Grandma Rose wanted me on the couch in the den, Grandpa insisted I share their more comfortable bed. He had a strong musky scent, and when I draped my arm over his, I could feel the mound of his powerful biceps. Now his chest emitted a sour smell and the muscles of his long arms lacked their former bold curves.

I watched Dr. Mathers's exquisite neurological examination: cranial nerves; tendon reflexes; and sensory perception of light touch, vibration, pinprick, and position. Dr. Mathers had to repeat his instructions several times in order for Grandpa to cooperate in coordination testing, essential in the assessment of the cerebellum. But when he did move his finger to his nose with his eyes closed and then walked in tandem heel to toe like a circus performer, each of these tasks was completed without faltering.

Dr. Mathers sat at the desk completing his note. I glanced at the paper. The golden nib of the pen traced a tight, precise script.

"It's an interesting case," Dr. Mathers said to me, ignoring my mother and Maurice as they helped my grandfather dress. "Did you notice the inconsistencies in distant memory as well as recent memory?"

I shifted on the low stool uncomfortably and nodded tensely.

"That's unusual. Seen in less than 10 percent of cases. Generally, recent memory is more profoundly erased and distant memory better preserved."

I replied I hadn't known the percentage was that low.

"To certify him with Alzheimer's for the nursing home," Dr. Mathers said, turning to my mother, "I need to bring him in for more tests."

"Alzheimer's for sure?" I interjected. "No other diagnostic possibilities?"

"Perhaps," Dr. Mathers replied. "The tests will show."

The doctor opened the lower drawer of the desk, searching for a hospital admission form.

"It seems like a nice home," my mother said.

"Don't know it," the doctor answered.

"They have RNs on site and a physician on call."

Dr. Mathers did not respond and began filling out the paperwork.

"What tests are you going to run?" my mother asked.

"Standard ones, and a few for interest."

"My husband and I have an appointment out of the city tomorrow," my mother continued. "In Connecticut. It's for his business. I could cancel it if I need to. How long will my father be in?"

"Two to three days," Dr. Mathers said, signing the bottom of the admission sheet. "Call me tomorrow in the afternoon. We'll finalize the discharge then."

He handed my mother his card, stood up, smoothed the creases on his white coat, and opened the door for us to leave.

After stopping at the admissions office and signing a guardianship form, my mother, Maurice, and I escorted Grandpa to the ward. A portly middle-aged nurse with a Jamaican accent, Mrs. Hendricks, greeted us warmly. She settled Grandpa into his bed, placing a pitcher of ice water on the night table. Then she took his vital signs.

"What is your favorite hobby, Mr. Sherman?" Mrs. Hendricks asked brightly.

"Fishing," he said. "Mostly flounder. Blues when they're running."

"Oh, my father was a fisherman, too," the nurse replied. "There is nothing more delicious than fresh fish right off the boat."

My grandfather's eyes widened.

"I'm off to Montauk in the morning," he told her.

I returned to Boston that afternoon and went for a long swim. I wanted to focus my mind in the submerged quiet.

The dementia, I now realized, had been evolving over several years. It was believed at first to be the appropriate sadness after Grandma's death, then excused as the deepening eccentricity of a famously rambunctious man. Although Alzheimer's disease loomed as the diagnosis, I was not convinced. Other disorders could cause chronic dementia: benign tumors of the frontal lobes of the brain; subdural hematoma (meaning blood collected between the skull and brain); metabolic changes from confused use of medicines or from excess levels of magnesium and calcium; heavy metals, like lead, or other toxins in the water or food; liver or kidney dysfunction; nutritional deficiencies of B vitamins; low thyroid function. Each could be corrected.

Specialists as eminent as Dr. Edward Mathers made no assumptions. They earned their place among their colleagues by identifying the arcane and not glibly endorsing the obvious. Although Dr. Mathers's demeanor was cool, I took that as his refined professionalism and the analytical persona of an academic neurologist.

Later that evening in our apartment in Boston, my wife, Pam, listened to the salient points of the consultation.

"You'll have to see what Grandpa's tests show over the next two days," she said gently. "There could be something reversible."

I grew that hope in my mind the next day.

Dr. Mathers here. Turns out to be hormonal, an abnormality in your grandfather's thyroid levels. You can't imagine how often I lecture the interns not to overlook chronic hypothyroidism in the evaluation of dementia or depression. We've started to correct the levels. I've asked the chief of endocrinology to see your grandfather about follow-up to adjust his doses of thyroid hormone. The dementia will resolve.

Dr. Mathers here. Well, we found a large but benign tumor in the frontal lobes on the brain scan. It was a surprise, since there were no signs of it on his neurological examination. Surgery is scheduled for tomorrow. You know what an outstanding neurosurgical service we have here at Columbia. The prospects of removing the tumor and fully restoring your grandfather's cognition are excellent.

My heart lightened as I created these scenarios in my mind.

"Phone call, Jerry," David Golden, a researcher in the lab, called out. I hadn't heard the phone ring.

"He's being discharged," my mother said in a panicked voice.

"What? When?"

"Now. Today. At the end of the day."

I heard the rough rumble of cars in the background.

"Where are you?"

"At a pay phone. In a gas station off the turnpike in Connecticut."

My mother had accompanied Maurice to his business appointment in Bridgeport. She had called Dr. Mathers's office, as instructed, for an update on my grandfather's evaluation.

"He said all the tests were done, that he got all the data he needs. They're sending Grandpa to the assisted-living home."

It didn't make sense. The discharge wasn't scheduled for two days. My mother had planned to take my grandfather from the neurology ward and accompany him to the home.

A sharp metallic click cut off her voice. There was a momentary silence, then a recording: "Please deposit another fifty cents for the next two minutes."

The line clicked again and I heard the loud grunt of a truck changing lanes.

"Maurice, two quarters. I need two quarters."

"Please deposit another fifty cents for the next two minutes."

I heard my mother's panicked demand for two quarters again. Then the line went dead.

Dr. Mathers had given his card only to my mother. I dialed by heart the hospital's general number.

"Dr. Edward Mathers, Neurological Institute, please."

"Clinic or administrative office?"

I thought for a moment and asked for the office.

I was greeted by the polite voice of a young woman. "He's still in clinic, Dr. Groopman. If it's urgent, I can connect you."

It was.

"Dr. Edward Mathers here."

"Dr. Mathers, this is Dr. Groopman. I'm back in Boston and—"

Before I could continue, he began to speak.

"We finished the testing in one day. Essentially consistent with Alzheimer's disease. Brain scan shows atrophy of the cortex. The spinal tap had normal opening pressures, the glucose was fine, no elevated protein, no cells. We put him through detailed pattern recognition exercises—you know, copying geometric shapes of increasing complexity. He had difficulty, as expected. Further memory evaluation was most intriguing. His fluctuating degrees of recent and past memory were a bit of a curveball. I've made a note of it in the record. But as I said, we're seeing this variation in about 10 percent of the cases we're analyzing. We'll write up this aspect of our series for publication soon."

"I just got a panicked call from my mother," I said, unable to assimilate the cascade of clinical information. "He's being discharged to the home today?"

"He is," Dr. Mathers replied dryly. "The home was contacted by my office and is ready to receive him. I need his ward bed. Another case is coming in for testing."

"But my mother planned to accompany him. She wants to be there when he enters the home. It's a strange environment."

"He's already en route," Dr. Mathers replied matter-of-factly. "He'll get there safely."

I hung up the phone and sat at my desk. *Dr. Mathers doesn't give a damn. He got what he wanted: more data for his research.*

I imagined my grandfather staring blankly from the window of a van, driven by a man he did not know, destined for a home that was not his.

My mother called about ten minutes later from the same station along the Connecticut Turnpike. Her voice quavered, and I asked if she was okay. Mom said it was cold, that the sun was setting. I informed her of what I'd learned from Dr. Mathers. There was nothing I could do.

"We'll drive directly to the home," my mother said forcefully, "and get him settled as best we can."

They arrived in two hours. My mother later told me she found Grandpa sitting in the lobby, a nurse by his side. He stubbornly refused to go to his room. During the ride to the home, he had become agitated and picked apart the threads of his shirt buttons.

My mother hugged him tightly, tears in her eyes.

"So this is the right hotel, Muriel?"

The Centerville Home was a renovated single-story country house set back from the main road on three acres of land. Each resident had his or her own room and toilet. The shower and bath were shared, situated at the end of the floor, and spotless. There was a sitting room off from the main entrance with a large beige sofa, upholstered club chairs, tables and lamps. All the levels were carpeted, and there were framed Audubon prints on the walls. At mealtime, the residents ate in small groups at round cloth-covered tables, and special kosher diets, like my grandfather's, were honored.

The staff on the premises included a manager with a degree in social work, a registered nurse to administer medication, an occupational therapist, and a games director who organized activities. All residents were examined monthly by a local physician, a specialist in geriatrics. It was costly, but there was no question that Grandpa's savings should be spent for his care.

"He's doing well there," my mother affirmed two months later.

He tied for first place in the checkers tournament. The staff also organized competitions to identify operas, and Grandpa proudly showed off his knowledge.

"And he has girlfriends," my mother said with a chuckle.

Each time she visited, he was holding the hand of another of the home's female residents.

"He calls them all Rose," my mother said, her voice growing lower. "He can't remember your name. Or sometimes mine."

***

"Open the door, Gramps," I said loudly. "Open up. It's Jerry."

I had come with my mother and Maurice for what I hoped would be a pleasant visit. I heard more furniture being pushed against the door.

"He's barricaded himself since this morning," the attendant said.

My mother stood to the side, gripping Maurice's hand. I pushed hard with my shoulder against the door. The barricade quaked, and the door began to give way. A sliver of light came from my grandfather's room. I pushed harder, the full weight of my body centered against the door.

Where does he get his strength from? I thought.

"It's Jerry, Gramps. It's Jerry."

The barricade finally broke, and the door creaked halfway open. I entered. The acrid smell of urine assaulted my nostrils. My grandfather was crouched at the far wall, his light hair matted on his bowed head. His dresser drawers were opened and their contents strewn on the floor: socks, underpants, dentures. His opera records were strewn and cracked. The handle of his phonograph dangled like a broken limb from the side of its platform.

I approached him slowly, speaking softly, like a trainer moving toward a cornered lion. He stood up and lifted his still powerful fists over his head. I froze in my tracks.

"It's Jerry, Gramps."

I steadied my gaze into his, willing some remote corner of his brain to recognize me.

"Remember? Jerry."

His steely eyes softened, and his clenched fists opened.

"Where is Rose? Where is Rose?"

Through that year of 1981, the staff at the home tried their best. They were not structured to administer differing doses of medication or to handle residents who required long-term restraints.

My mother persisted in the hope that these episodes of aggression would wane, as they had after the attack on Uncle Sam. She visited frequently and each time followed the instructions of the caregivers: always reorient him when he is confused; repeat your name; tell him where he is; specify the day of the week, the month, the year. But her efforts were fruitless.

I had left Boston the year before to become an assistant professor at UCLA. Los Angeles was a welcome change. Pam and I lived in a small house perched on a hill overlooking the Pacific. There were citrus trees in the yard, and at night, the howl of coyotes. I felt untethered as I watched the ocean shimmer while the sun turned its back on the rest of the country and moved beyond the limits of land.

In June of that year, Pam was pregnant with our first child and travel became problematic for her. So I flew alone to the East Coast when I heard my mother's predicament.

My mother was waiting at the gate. She cried when I kissed her and then asked after Pam. I assured her that all was well. We walked together from the TWA baggage terminal to the parking lot. Maurice greeted me with a warm handshake and insisted I sit in the front seat of his Oldsmobile for the ride out to my grandfather.

I steeled myself as I walked down the carpeted hallway to his room. A young male attendant with a gentle Irish brogue opened the door with a key and then stood discreetly to the side.

Grandpa was seated in a large chair. His hair was neatly combed, his face shaven, his pajamas clean and neat. When I moved toward him, he grunted the deep visceral sound of a wounded animal. He pulled forcefully against the tight straps that bound his arms to the chair. But the restraints held.

"Gramps, it's me, Jerry."

My mother bit down on her lip. Maurice reached for her hand and drew her close to him.

My grandfather looked wildly at me.

"Let me up," he yelled. "Let me up, you bastard."

I walked closer, my hand extended, palm open.

"I'll kill you. I'll kill you all with my bare hands."

Grandpa lifted his head and growled. A glob of warm saliva hit my cheek. I withdrew my hand and turned away.

My mother, Maurice, and I sat with the director of the home in her well-appointed office. She was a middle-aged woman, slim, with glasses and a gentle mien. Her mahogany desk held photographs of her family. Grandpa's file rested in front of her.

"I am sorry," she said. "He's beyond our level of care."

The violent spirit that had burst forth those years ago when Grandpa had threatened Uncle Sam had reemerged. And as in his beloved novel by Robert Louis Stevenson, in the end there was no potion to transform the monstrous Mr. Hyde back into the humane Dr. Jekyll.

"We'll pay for the property he destroyed," my mother offered. Grandpa smashed the furniture in his room. He was found defecating in the hallway. Yesterday, he had pulled the sink in his bathroom off its backing.

The supervisor declined, saying it was not necessary. The home had insurance, and none of the other residents had been hurt.

We drove the short distance to Maurice's home. No one spoke.

Penny Goldman, a close friend of my mother's from the time they were schoolgirls in the Bronx, was waiting for us. She lived on Long Island, and when my mother married and moved to Maurice's house, the relationship was rekindled.

Mom prepared tea and set out my favorite poppy-seed cookies on a glass plate. I took one but had little appetite.

"There is a downstairs den where he could live," my mother said.

I glanced at Maurice. His expression was impassive.

"You have to put him away," Penny Goldman said.

My mother's face tightened.

"Muriel, it's awful. I went through it all with my mother two years ago."

Mrs. Goldman took another sip of tea. She had put her mother in a state institution and no longer visited her.

"Muriel, you have to face the fact that he's gone. The man you see is not your father—he's someone . . . else. I tried to hang on, too. But I finally accepted that my mother is dead even though her heart is still beating."

I listened silently.

"You feel guilty. You feel like you're turning your back on them. I went to a therapist for help, I felt so lousy about what I did. He helped me accept reality: my visits to my mother were useless. They just upset me—she never knew who I was, where she was."

My mother exhaled with agitation.

"And, bluntly speaking, it's a waste of money to put Max in a private facility. A state institution takes Medicare. You'd do better to keep whatever is left of his savings for the family."

My mother looked down at the cup of tea before her.

I reached out and gripped her trembling hand.

"I want to care for him," my mother said weakly. "I'm not a nurse, but if he was on the right medication maybe he would be . . ." Her voice trailed off. "Maybe he would be better."

"There is no 'better,' Muriel."

I knew Mrs. Goldman was right in that respect. The elaborate relays of nerves that once connected within Grandpa's cortex were deformed into frayed and twisted tangles. Where these vital neurons once traced paths of love and joy and humor, there now were lifeless tracks, like fossil imprints. A scan would show the gross outcome of the relentless decay of his cells, his cerebrum further shrunken and atrophied. And without the restraint of this upper cortex, there emerged a primitive brain from below, like a fanged serpent disturbed under a rock, lashing out with instinctual fury.

Would this be my fate also? I had inherited his height, the width of his shoulders, the line of his eyes. Had he also passed to me the destructive seed that flowered in his cranium? Would its poisonous roots one day strangle my identity?

Was he really dead?

Mrs. Goldman left.

My mother busied herself preparing another pot of tea. After I began the second cup, Mom asked what I thought she should do.

I first stated that it was impossible for Grandpa to live with them. The burden was too great.

My mother reminded me that Grandma Rose and Uncle Sam had cared for my great-grandmother Miriam through the twilight years of her senility. She had been bathed and fed in a large hospital bed occupying the study on the first floor of their Bronx home, and she died there in her late nineties. A nurse had come to help as needed.

That was different, I countered. Not just a different time but a different clinical situation. My great-grandmother's senility made her passive and involuted, not abusive and violent.

I did not say that my grandfather's needs would strain the growing fabric of my mother's new marriage, that her new life, and Maurice's life, would be stunted by his presence. I knew my mother's love for her father was fierce, and any suggestion that he was secondary would be rejected out of hand.

After an hour of discussion, I finally developed the argument that swayed her.

"It's not so much you and Maurice that I'm worried about," I said, "even though I do worry about you both. I worry about Grandpa. He can't get the care he needs if he's here. His case is too medically complex. He needs constant professional attention, nurses and doctors. He's being started on Haldol for the aggression. Decisions will have to be made daily about the dose or adding other sedative medications. He'll get urinary infections from the catheter. His skin requires attention to prevent bedsores. You don't have that expertise. It's unfair to *him* to live with you."

My mother closed her weary eyes.

Maurice repeated his feeling that it was her decision and hers alone.

The private facility was a four-story brick building located near my mother's town on Long Island. It housed about a hundred patients. The lobby was not decorated, and the walls of the resident floors were barren. The building had the pungent smell of ammonia. Grandpa was in a small single room. Most of the day, he lay in a hospital bed in diapers, tightly restrained.

"But he's kept clean. I feed him some pudding when we come, if he's awake enough."

A doctor saw him weekly. Grandpa received high doses of sedatives in addition to the Haldol to stifle the violent urges that came during his alert state. Without warning, my mother reported, his face would twist and writhe in spasms, a side effect of the medication termed "tardive dyskinesia." Saliva flowed freely from his lips onto his chin. His hands trembled when he tried to hold a spoon, spilling the food on his chest. His high cheeks became sunken, his eyes retracted in their sockets. Under the pale sheen of his skin were the knobby outlines of his skeleton.

On one visit, my mother told me, the attendant gave permission to undo his restraints. My mom and Maurice tried to take him for a walk, but he faltered and returned to bed. His powerful limbs had shriveled into useless sticks. He hardly had the force to hold up his head.

On March 19, 1983, my mother received a telephone call. Grandpa had pneumonia and was transferred to a local hospital. She and Maurice went there. He had no IV, was given no antibiotics. My mother understood and assented.

The next day, what remained of his life was gone.

⌒

I closely follow scientific developments on Alzheimer's disease. Researchers have analyzed brain tissues at autopsy and found unusual proteins in the signature tangles and plaques. But how these clusters of proteins arise remains obscure. Families with many affected members have been studied seeking an aberrant gene; none has yet been identified that explains the disorder. Many hypotheses about an environmental cause of the illness have been aired, ranging from excess aluminum intake in the diet to covert viruses. But, to date, the genesis of Alzheimer's disease remains a mystery.

Several articles on what might help prevent the disease caught my eye. The data were retrospective, drawn from comparing patients with Alzheimer's to people free of the disease. People who regularly took nonsteroidal anti-inflammatory medications, like naprosyn and ibuprofen, had a markedly lower incidence of Alzheimer's disease. The researchers postulated that an enzyme in the brain, called COX-2, which can cause neurons to die, is somehow triggered in Alzheimer's disease. Blocking COX-2 with the nonsteroidal anti-inflammatory drugs protects the brain cells from untimely death.

The data from these studies could be artifacts of retrospective analysis. A prospective study, assigning patients to a new COX-2 blocker, celexocib, or to a placebo, is underway to obtain definitive proof. The drug has few known side effects and can be taken daily without interruption.

I was prescribed the drug for my chronic back problem, and its potent anti-inflammatory properties soothed my frequent aches and pains. But I was particularly welcoming of its still unproven potential. I am haunted by the experience of my grandfather and anxious to do something should I be programmed to his fate.

The sixteenth year since Grandpa Max's death neared as I wrote this chapter. I took the opportunity to revisit the events of his life with my mom. We were warmed by the memories of his rambunctious exploits. We laughed about the time he and his fishing buddies marched into our new house in Queens in search of beer after a Sunday expedition to Montauk. Mud and fish slime flowed onto the pristine pastel blue carpet. My mother demanded their catch as compensation for the mess they made. And she got it.

Then we turned to the painful memories. I asked Mom what might have made it easier.

"If someone had told me what to expect."

She accepted that the disease is insidious in onset, and that its course is variable. Grandpa had had a very good year at the Centerville Home after the violent episode with Uncle Sam. But no one took the time to prepare her for what would likely follow.

Mom's voice grew sharp as she talked about Dr. Mathers. She understood that there was no treatment once the diagnosis was confirmed. But, from that point on, the doctors and other caregivers should realize that what was done was as much for the family as for the afflicted. Dr. Mathers's cold indifference to her simple request to accompany Grandpa to the home could not be forgiven. My mother blamed herself for not being more alert to his agenda. I should have been as well.

And Penny Goldman's advice? Did it really matter where Grandpa was those last six months and if she visited?

"I thought about it a lot," Mom said. "There were times Maurice and I visited when there was no expression on his face. He was like a vegetable. But a few times we went it was different. True, he didn't know who we were, didn't know his own name. But our being there brought a smile to his face. He could still feel pleasure. There was a person in the bed."

Decoding Destiny

Not long ago, Karen Belz made an appointment to see me in my office. I'd known her for years, as a family friend. She worked as an English teacher at a local high school, played in the same tennis league as my wife, had two school-age children, and was active in town politics. Always frank and direct in her dealings, she expected the same from others. Now she was telling me that she wanted to be tested for mutations in the so-called breast cancer genes, BRCA1 and BRCA2. "I can't continue in this limbo," she said. "I have to know."

I wasn't surprised, because she wasn't the first member of the Belz family to visit my office. Karen's mother, Eva, after a long struggle, had succumbed to breast cancer some three years before. Five months earlier, Karen's sister, Ruth, only two years older than she was, had developed the same malignancy, at age thirty-six. By the time Ruth's gynecologist detected the cancer on her routine surveillance mammogram it had spread to her vertebrae and her lungs. I was currently treating her with intensive chemotherapy and radiation. Shortly after her tumor was diagnosed, Ruth had tested positive for a BRCA mutation linked to breast cancer and had pressed Karen to be tested as well.

Marek Belz, Karen and Ruth's father, was strongly opposed to his daughters' being tested, fearing that the result would one day be exploited to harm his family. He was familiar with the abuse of genetics. Marek had escaped from Poland a year before the Nazi invasion, with Eva. When she became ill, neither could tell me about the occurrence of cancer in close family members: every member of their own generation and every member of the generation before had perished in the Holocaust. Karen, however, was unmoved by her father's opposition: she told him she'd simply have to take the risk. Since Ruth's diagnosis—first of cancer and then of the aberrant gene—the anxiety of not knowing had dominated Karen's thoughts, preventing her from getting to sleep and then shocking her awake at 3 or 4 A.M. Her daydreams were penetrated by memories of her mother's slow death, and she had nightmares that Ruth would soon be lost as well. Not knowing seemed worse than knowing.

All participants in cancer-genetic-susceptibility programs are required to read and sign an informed-consent document before being tested. At Beth Israel Deaconess Medical Center in Boston, ours comes with some assurances. The patient can elect not to know the test result immediately but to wait until she is ready, or even change her mind and never know; the result is to be given only to her, in person and by her own physician; the fact that she has been tested, as well as the test's outcome, does not appear in her medical records but is stored in a locked research file, identified by a code number; the list connecting the code with her name is kept separate from the test results; and no insurance company, governmental agency, health worker, family member, or other party can obtain the test result without her written permission.

I informed Karen of the protocol, and read to her a section entitled "Potential Risks of Knowing Your Genetic Test Result," which states, "This information may cause you distress, sadness, depression, anxiety, or anger."

"I look forward to all of the above," Karen said acidly.

BRCA testing is one of the first fruits of the human-genome project. The project is vast and ambitious, and has enlisted the energies of scientists and clinicians around the globe. The goal is to decipher the code of all human DNA—our genetic blueprint—and then to use the information to understand how each gene contributes to health or to disease. The task should be completed within one or two decades.

The informed-consent document that Karen signed before being tested is specific for the genetic testing of BRCA1 and 2, two genes among the estimated hundred thousand genes contained within our forty-six chromosomes. All human beings carry two copies of each of their genes, one inherited from their mother and one from their father. Women who inherit a single defective copy of either BRCA1 or BRCA2 are at a significantly increased risk for breast and ovarian cancer over the course of their lives. Of some two hundred thousand new cases of these two cancers that will occur in the United States this year, between 3½ and 7 percent will be the result of inherited mutations in the BRCA genes. There is a particularly high incidence among Ashkenazi Jews, like the Belzes. This probably reflects the so-called founder effect: the Ashkenazi population arose from relatively few "founder" families, which migrated east into Poland, Lithuania, and Russia, with very little marrying outside the community, and thus conserved the mutation from generation to generation. The combined frequency of BRCA1 and 2 mutations among Ashkenazi Jews is greater than 2 percent, a very high number in population genetics.

"Why would a gene that promotes deadly cancers be conserved during evolution?" Karen asked.

There is no exact answer, I told her, but we could extrapolate from Darwinian principles. Nature cares only about an organism's

reaching the point of reproduction and nurturing the next generation. Since the mutated BRCA genes do not impair fertility, and
generally wreak their havoc beyond the age at which children would
be produced and reared, there is no apparent evolutionary pressure
to select against the genes. Their perpetuation might even be a benefit to the population, by shortening the life span of adults who were
taking precious food and water away from the younger generation.

"What a wonderful Malthusian gift," she said with a grim smile.

Mutations often occur within our genes, for the enzymes that
copy DNA when a cell grows and divides are imperfect and make
random mistakes. Mutations in DNA also occur when we are exposed to radiation or to certain toxins in the environment. In most
instances, these mutations don't really matter: either the changes
are trivial and well tolerated by the cell or they're so damaging that
the cell dies, with limited consequences. But some mutations are
neither trivial nor lethal. These pervert the cell's behavior, causing
it to proliferate wildly and assault vital organs. This is the pathological process we call cancer.

The normal BRCA2 gene is composed of 10,254 nucleotides, or
DNA building blocks. The BRCA2 gene carried by the Belz family
was missing nucleotide No. 6,174. This single omission results in a
short and crippled form of the protein that BRCA2 codes for. At the
time of the test, I had explained to Karen that the normal function
of BRCA genes is understood only in general terms. The most recent research suggests that they somehow repair damaged DNA, re-
habilitating nascent cancer cells like an interventional therapist.
Why, when the genes malfunction, cancers arise only in breast and
ovary, and not in other tissues, is still a mystery, but it suggests that
sex hormones like estrogen somehow modulate its activities.

Five weeks later, Karen and I sat again in my office. I had just given
her the test result. "So I'm destined for cancer, just like my mother

and sister," she said. "Talk to me straight, Jerry. I should get rid of my breasts and ovaries, shouldn't I?"

Karen's style was to cut to the heart of the matter, but I worried that she was moving too fast. It wasn't possible to say even that she was destined for cancer, let alone talk to her straight about what to do. I felt off balance, without a platform of established clinical data upon which to stand. It is a moment of inadequacy that every physician dreads, realizing that while current knowledge is severely limited, a course must be taken—a course that might prove fatally flawed in the future. I paused, studying Karen's unflinching face: deep-set almond brown eyes, sculpted Slavic cheekbones, and determined chin.

I was tempted to hide behind the popular notion that patients alone are "empowered" to choose their therapy, and simply tell her that it was her decision. But wasn't that a kind of cowardice—an easy out? The decision had to be hers; but that didn't mean she had to make it alone. "Remember that our genes are just the starting point, far from the whole picture," I cautioned. "Their program during life is modified by many factors. Despite having the same mutation in a BRCA gene, you are different from your mother and from Ruth."

"But how different? That's what I need to know. And you can't really say, can you?"

"I can't," I admitted. I went on to explain that the BRCA2 gene was identified in 1995, just two years before, so the available data tell us mainly about the overall susceptibility of the affected group to cancer and not about specific individual outcomes. One set of clinical data on BRCA mutations indicates that the age at which tumors first occur is highly variable. Although recent surveys of Ashkenazi Jews indicate that the risk for breast cancer with a BRCA mutation is about 60 percent, in families with several closely related cases of the disease the risk is higher, on the order of between 75 and 90 percent. Still significant, but somewhat lower, is the risk for ovarian cancer, about 20 percent. "Keep in mind that these odds are ac-tuarial—projections over the course of a person's lifetime, culled from a handful of preliminary studies in the United States and Is-rael," I told her. "Nonetheless, I think one can accept them as accu-rate approximations."

I stopped speaking, in part to give Karen a chance to ask ques-tions, in part because I didn't like the way I sounded. It was impor-tant to provide her with all the available information on the mutation, but I had heard my language becoming increasingly re-mote and clinical.

Karen's face had turned into a mask, and I feared that she was numb from having learned her test result, that the swirl of uncertain numbers had passed by her in a blur. When I asked if everything was clear, however, Karen said it was. She had already garnered most of this information from the Internet. She was waiting for me to get to her question.

"Should you have your breasts and ovaries removed?" The ques-tion hung suspended between us for a long moment. "To answer most helpfully, I'd have to be sitting where you are, rather than on my side of the desk, and be a thirty-four-year-old woman, with two children and a solid marriage. So I'm limited on these counts. But if I had a genetic mutation analogous to yours, carrying as much as a 90 percent risk of, say, testicular cancer, and a 20 percent risk of prostate cancer, and my father had died from it, and my brother now had it, despite close medical surveillance—regular physical and radiological examinations, biopsies, and blood tests—and he was undergoing intensive chemotherapy and radiation treatments for metastases, with all their toxicities and an uncertain chance of cure, what would I do?" I wanted, as I spoke, to spark intuition, hers and mine. "And I faced two unsatisfactory choices: one conservative, waiting under surveillance, and one radical, undergoing surgery, with no real middle ground—"

"Is there actually nothing more than that?" Karen interrupted. "I accept that there's no simple solution. But you feel desperate when you hear that that's all there is. You want to do something for yourself, something proactive, and not just live under surveillance. It seems like passively waiting in the ghetto to be selected."

When Ruth became ill, Karen told me, well-meaning friends had bombarded them both with suggestions: eat strictly organic, take megavitamins, avoid all alcohol and fats, try special herbal tonics and Eastern healing techniques. Karen and I now talked about such suggestions. Many of these nostrums were harmless, and some, for all anyone knew, might even be helpful, but they remained unproved and could not be relied on to prevent breast or ovarian cancer in a woman who was genetically predisposed. Prophylactic studies of estrogen antagonists like tamoxifen were just being organized and would take years to yield data on risks and benefits.

As we spoke, I could see her brow fall and her eyes become heavy. I knew that what I had to say would be even more unnerving and had best be delivered quickly.

"If surgery were to give me a full guarantee against cancer, I'd probably sacrifice intimate parts of my body in exchange for it, with all that that means regarding sex and self-image," I said. "But no one really knows how much protection is gained by such mutilating surgery, because the data are scant." No one had had the opportunity to follow women with BRCA1 or BRCA2 mutations who chose surveillance instead of having their breasts and ovaries removed. There were only extrapolations from the past, when healthy women who had strong family histories of these cancers chose surgery. But this was a heterogeneous group, not genetically defined, and had come into being before the era of high-quality mammograms, pelvic ultrasonography, and blood tests, which could identify tumors at their onset. "Unfortunately," I added, "it's all we have to go on."

"And?" Karen put in.

"And removing both breasts is roughly estimated to reduce your risk of cancer by 90 percent but not to zero," I said. Even the most meticulous mastectomy, I explained, leaves some cells behind that can later transform and become malignant. Oophorectomy—removing the ovaries—would be likely to reduce her risk of cancer, but, again, not to zero. There are often microscopic islands of ovarian cells growing outside the ovaries. These islands stud the peritoneum, the lining of the abdomen, and are too small to be found and cut out by the surgeon during the operation. They can provide the seeds of cancer after the ovaries are excised.

"And the success of intensive surveillance in finding the cancer early, before it spreads, like Ruthie's?" Karen pressed.

"Unknown over the long term," I replied.

I paused, then told her that nonetheless most women I care for, and presumably most women nationwide, currently elect this conservative approach of surveillance rather than the radical surgery.

In the shared silence, Karen gathered herself into a tightly held coil, as if she were a child chilled to the bone and were trying to protect her body's ebbing warmth: head bowed, arms crossed over her chest, legs tightly drawn together. "I can't do it," she said finally. "I can't get rid of my breasts and my ovaries."

I was taken aback. As we talked, I had been imagining Karen as my wife, Pam, and the Belz family as my in-laws, and I had convinced myself that I would want Pam to have the operation—that the physical alteration of her body would not change who she was or the substance of our relationship. And Karen was a friend as well as a patient. I had mourned the death of her mother and was now afraid that Ruth would soon succumb as well. Above all, I wanted Karen to be saved. Surely that imperative overrode the loss of breasts and ovaries, of self-image and libido. But I hadn't understood her as well as I thought I had.

"The final decision shouldn't be made today," I said, giving us both more time. "Just as we didn't immediately go ahead with the test. It's best to think more before committing to a course." I watched her carefully. She exhaled heavily and nodded.

We agreed to meet in a week to talk again about the options. I stated that afterward I would refer her to the clinical psychologist in the cancer program, to give her the opportunity to explore her decision with another professional, and perhaps from a different perspective.

And maybe I needed a different perspective myself. After Karen left, I tried to organize my thoughts by recording a summary of our conversation in her medical chart. But I was stopped by the realization of something that neither of us had brought up. What of her children? Her daughter was thirteen and her son eleven, both at an age where explanations are required. When and how should they learn about the BRCA2 gene?

My focus was drawn to the pictures of my family arrayed on my desk. There were Pam and I dancing at our wedding, joined by a white handkerchief, in traditional Ashkenazi style. Pam's family had a history of breast cancer, and this past year my mother had developed it. Beside the wedding photograph was a picture of Emily, my five-year-old daughter, on a swing, caught by the camera in a moment of fearless glee. What terrible aberrations hid in the fabric of her DNA, waiting for age and hormones and the myriad triggers of the environment to unleash them? Would the effort to unravel DNA condemn Emily to the twilit terror that Karen had just entered? Perhaps it was best for all of us to remain ignorant, so that life could progress naturally, without the burden of deadly prophecies. It sometimes seemed as if the decoding of our genome would cause a fundamental change in how we perceive time—as if we would come to ponder not the infinite time of an expanding universe but the sharply limited span of our existence. Like Karen, all of us will face the choice of learning our probabilities of illness. In addition to those in BRCA1 and 2, genetic mutations that predispose people to Alzheimer's disease, colon cancer, Huntington's disease, endocrine tumors, and melanoma have been identified. The list will grow until it encompasses all our potential pathologies. We might try to shrug off the knowledge, or run from it, but when we had quiet moments during the day or woke in the middle of the night we would be forced to accept it as our constant companion, because we could see its features in our very being.

And then it struck me that these new genetic terrors were like the ones I had come to know in my years of clinical and scientific work on AIDS. People with HIV permanently acquire the destructive genes of this virus. They, too, live with a genetic time bomb. Vociferous debates racked the at-risk communities, particularly those of gay men and hemophiliacs: Should you be tested? Could the results be used to deny you insurance or opportunity in the workplace? What could you do for yourself to stave off the disease? Why did medical science seem powerless to find a cure?

A few years back, I had faced the decision of whether or not to be tested for HIV. All health-care workers were encouraged to have themselves screened, and so were researchers who, like me, had worked with HIV in their laboratories. I knew I was at especially great risk of having been exposed to the virus. Before gloves and other precautions were instituted in hospitals, my hands had been soiled directly by infected blood and secretions, and I'd once pricked my finger deeply with a contaminated needle after performing a bone marrow biopsy on an AIDS patient. I had also received multiple blood transfusions during spine surgery in 1980, a time before blood products were screened for HIV.

For months I avoided scheduling the test, finding excuses in the demands of work and travel. But finally I went ahead, feeling, like Karen, that I had to know. I also accepted the fact that the result could change my plans for work and family. Even after obtaining the negative result, I was still visited by the nightmares I had experienced while waiting, and found myself wondering whether I might be one of those rare people whose infection was not detected by the test.

Though I observed the anguish among those who learned that they had tested positive, I was also struck by the way the awareness could catalyze beneficial changes. It led many people to take better care of themselves—stop smoking, end drug use, improve diet, take up exercise, seek stress reduction. On a community level, activist movements forced the FDA, universities, and pharmaceutical companies to accelerate the process of drug research and development. Above all, the knowledge that one carried HIV and was mortal gave many young people a precocious sense of maturity and wisdom.

The advances that had been made in developing safer and more effective treatments for HIV could not have occurred without testing people with the virus at different stages of the illness. The most important breakthrough—the creation of the protease inhibitors—grew from an intimate understanding of the genes of HIV and of how they functioned in an infected host. If we had not used our molecular tools to test for the virus and assess the range and the patterns of its growth in people, we would not have developed these potent drugs.

Testing also revealed that some people never became infected despite extensive exposure to HIV. Many of them had a mutation in a host gene that conferred resistance to HIV. Such people are serendipitously resistant to the virus, because HIV is unable to get a grip on their blood cells and penetrate them. The inherited mutation causes no apparent harm: these fortunate individuals seem completely healthy.

Such findings provide the intellectual framework for the design of therapies that mimic the serendipitous mutation and confer protection against the virus. Again, if people had not agreed to be tested and offered themselves as subjects for study, with no apparent benefit other than knowing they were contributing to scientific investigation, we could never have obtained this information. Commuting the death sentence of AIDS will be realized only if testing continues and more clinical data are obtained.

The ferment around breast cancer recalls the early days of AIDS: acute public awareness, increasing political pressure, the marshalling of research prowess and resources. It may take years or decades for researchers to come to understand the complex genetic underpinnings of breast cancer and to use this knowledge to design targeted ways of preventing or curing it. But that moment will arrive. And it will do so only as a result of the exploitation of clinical material: patients' blood cells and tissue biopsies and DNA.

So this was one reply to Karen's question of what she could do for herself—and, indeed, for her children if they, too, had inherited the gene. I planned to encourage her to be an active participant in the fight by using her political skills to push society harder and faster to address her needs. I would ask her to enter our ongoing study of the clinical and psychological condition of women who have tested positive, allowing us to collect data on her medical course and on her functioning in the home and in the workplace. Studies on women like Ruth—those who have a BRCA2 mutation and a family history of breast cancer—will be vital to understanding the origins of the disease. Just as important are the insights that can be gained from the exceptions to the rule, as we determine why at least one in ten women with a BRCA mutation does not progress to cancer.

For a week I anxiously anticipated Karen's return appointment. I had not been trained in medical school or during my clinical residency to handle the issues that confronted me. I had not been instructed by a senior attending physician in how to speak and listen, how to decide which questions are appropriate and which taboo, or how to make one's case for the best option.

Karen arrived promptly for her midday appointment. It came during the school lunch hour, and I noticed her sitting in the waiting room, grading papers and recording the scores in a large ledger she balanced on her lap. She greeted my secretary warmly and entered my office with a polite smile.

Karen spoke first, evenly and strongly. "I'm going to have my breasts and ovaries removed."

I was stunned by the reversal.

"No, I'm not losing it, Jerry. And I will see a psychologist to explain my decision. The shrink will find me compos mentis."

"What changed your mind?"

"Not 'what'—'who' should be your question."

Karen explained that when she returned home and told her husband, Sam, of our discussion, he couldn't believe she hadn't decided to have the surgery. He said that he loved her deeply, that she would remain desirable, and that their marriage was resilient. Even so, she was undeterred. "He was speaking theoretically, the way you were last week about your facing what I am," Karen continued. "I know that both of you were trying your best to help and comfort. But neither of you can really see yourself in my place."

I agreed, and then I saw who could. "Ruth," I said.

Karen nodded. I recalled telling her that she was different from Ruth. "How different?" had been Karen's reply. Now Karen said, "She called me several times after the appointment, but I repeatedly put her off. I couldn't bring myself to discuss it. She was always the big sister, scouting out danger and steering me away. This time, I thought I had to deal with it myself, without her. But Ruthie wouldn't let it go."

Ruth came to the house Sunday morning, Karen said, and Sam took the kids out to McDonald's. The two sisters sat at the kitchen table over untouched cups of black coffee, and they talked.

"Ruth said that I was one of the lucky ones," Karen went on, her eyes moist. "I had been given advance warning, by our mother and by her. That my relationship with Sam could survive any changes in my body, as hers had with her husband, despite the cancer and chemotherapy. And that I would never stop being who I am." Karen paused, and I reached out to grip her trembling hand. "Ruth said she wished someone had told her to have a mastectomy. Not just in hindsight, not just because now she has cancer. But because she would know that she had done everything possible to try to prevent it, and wouldn't live with regret." Almost in a whisper, Karen added, "I want so much to see my kids grow up. Ruthie doesn't think she'll see hers."

Karen and Ruth both entered clinical and epidemiological studies in 1997—studies that will help us understand how to use the genetic information we have. But clinical testing and observation are only part of the solution. The laboratory is where the tools to disarm mutant BRCA genes in the next century will ultimately be developed.

Such research could eventually permit us to use genes like drugs, as a form of therapy. This concept of gene therapy has great appeal: it is targeted to the fundamental cause of the cancer, in contrast with what is brutally termed "slash, burn, and poison," meaning surgery, radiation, and chemotherapy. If the issue is a mutant gene, then block it and substitute its normal counterpart. There are still many formidable obstacles to achieving this goal—obstacles that will take years to overcome. No technology exists yet to deliver the healthy gene throughout the breast tissue and the ovarian tissue, to switch off the aberrant BRCA gene and make sure that the healthy one functions properly. But recent successes in locally bypassing blocked arteries in atherosclerosis and restoring immune function in inherited immune deficiencies by the use of gene therapy indicate that one day the technology will be developed to help families like the Belzes.

One last thought came to mind as I contemplated the uncertainties that patients like Karen would be facing. It had to do not with science or society in the future but with wisdom from the past. Karen began each school year discussing Greek mythology, Bible stories, and fairy tales, showing how they contained enduring themes that were to be the subjects of her course in modern literature. Within the legend of King David was a parable concerning his desire to know about himself. After learning everything that was within his ken, David had beseeched God to reveal to him the date of his death, so he could plan his life accordingly. God had answered that no man can know the time of his passing. David was to live with an acute awareness of his mortality, and in this way more wisely fulfill his life.

I had taken this tale at face value for many years, assuming that it spoke of God's omniscience as hidden from man, and of the fixed limits we face in trying to learn life's secrets. But recently I had been wondering whether there was another interpretation. Perhaps God did not inform David because God did not know; perhaps the date and circumstances of David's end were not determined. The hope that sustains us comes from the belief that our future is yet to be created, and is created in part by us.

The Lottery

"The Irish don't run from a fight," James Leahey said in a determined voice. "And we're known for our luck. Don't write me off."

James was told by a surgeon to do nothing, that his cancer was too advanced. The bravado was to convince me to take on his case and battle the melanoma consuming his lungs, liver, and spleen.

I needed no convincing. A powerful new experimental treatment, gamma interferon, had just been developed. James *was* lucky—in the right place at the right time. I would give him gamma, I thought, and save his life.

It was two weeks earlier, on his thirty-fifth birthday, February 11, 1985, that James Leahey first noted the tumor. A tall, broad-shouldered man with slick raven hair, a ruddy complexion, and playful brown eyes, he liked to "dress sharp" for work as a claims clerk at Massachusetts Health and Life, a local insurance company. So he chose a tight-fitting dress shirt to look especially good on his birthday. But he couldn't close the topmost button. At first, James thought he had gained weight, yet the bathroom scale read the same as always: two hundred and nineteen pounds in his stocking feet. Then he felt the hard bulges in the flesh of his neck.

His primary care doctor found a small black mole on the crest of

James's left ear. It projected tentacles into the surrounding dermis. A biopsy of the lesion confirmed the presumptive diagnosis of melanoma. On physical examination, enlarged lymph nodes were present, not only in his neck but also under his arms and in his groin. A chest X ray showed multiple masses in his lungs. CAT scan revealed tumors in his liver and spleen but no spread of the cancer to his brain.

James Leahey hadn't noticed the irregular mole before, since there were many similar black and brown "beauty marks" on his body. Melanoma, he knew, was associated with exposure to ultraviolet light. He had spent many summer weekends at nearby Revere Beach and confessed he often forgot lotion.

"You don't think I should get a one-way ticket to Hawaii?" James asked with a nervous smile.

"I don't," I said, leveling my eyes into his, "and I'm a straight shooter. I'll never lie or sugarcoat a diagnosis. It's true past therapies have largely failed. At best, they shrink the cancer for a few months, and then it returns."

James's gaze faltered for a moment.

"We have a new experimental treatment, called gamma interferon. It looks more powerful than anything I know of. Clinical trials have just begun, so we don't have human data yet. But in mice with melanoma, it's a home run. The tumors just melt away . . . permanently."

James clasped his broad hands together.

"Plan on a round-trip ticket with this treatment."

I had longed to say this to a patient for some seven years. My first research project, as a fellow in oncology at UCLA in 1978, was on gamma interferon. I worked long days and nights with a Ph.D. biochemist to purify the protein. Gamma, as my laboratory chief put it, was the "Holy Grail" of cancer therapy. It would realize every oncologist's dream: stimulate the patient's immune system to attack and destroy a cancer, and make radiation and chemotherapy moot.

Interferons are naturally occurring proteins present in all of us. They were given their name because the molecules were first recognized as "interfering" with the growth of viruses. Only later were their anticancer properties discovered.

There are three types of interferon: alpha, beta, and gamma. Each is made in minute quantities in blood and other tissues, so it is difficult to obtain sufficient amounts for therapy. The advent of genetic engineering in the 1970s solved this dilemma. Scientists were able to isolate the gene for human alpha interferon. The human gene then was inserted into special bacteria. These genetically altered bacteria were grown in large vats, similar to those in which beer is brewed. The microbes produced vast quantities of human alpha interferon, which was readily purified.

Desperate patients with cancer flocked to the experimental trials of alpha interferon. The treatment had significant side effects: fever, muscle aches, anorexia, and fatigue. And despite very high doses, it failed to shrink the tumors. Only a rare form of leukemia, called hairy cell, and a rare form of skin cancer, Kaposi's sarcoma, yielded to alpha interferon.

Melanoma was included in the studies with alpha interferon. The cancer occasionally regressed with the treatment, but grudgingly and temporarily. At best, one in five patients showed partial shrinkage for a few months. Only one patient I treated, a middle-aged schoolteacher from New Hampshire, had had substantial regression of the cancer in her lymph nodes and liver. Her partial remission lasted seven months. The melanoma then grew back explosively, encasing her lungs and invading her brain. This experience resembled the "best" reported outcomes.

Clinicians who prided themselves as being "in the know" were not surprised. Their view was that the wrong kind of interferon had been pursued. Laboratory studies showed gamma to be much more potent than alpha in stimulating the T cells and macrophages of the immune system. I recall being dazzled by my own experiments on gamma at UCLA. Even the impure preparations we obtained turned these immune cells into merciless executioners of tumors. And the cancer that appeared most susceptible to gamma's guillotine was melanoma.

In the spring of 1983, three years before I met James Leahey, I had sat among hundreds of others scientists and clinicians at an interferon conference in San Francisco. Our team at UCLA was close to winning the gamma race. We had published an article in *Nature*, the prestigious scientific journal, on the partial purification of gamma and its astounding properties as an immune stimulant.

My pulse quickened as our main competitor, David Goeddel, a lanky, laconic molecular biologist at Genentech, a leading biotechnology company, ascended the podium. With no prelude, he switched on a slide showing the gamma interferon gene. There was an audible gasp in the auditorium. He had beaten us and every other lab. He had cloned gamma and his company had bacteria producing the protein in pure form. My work at UCLA would be, at best, a footnote.

I returned from California to Boston that July as an assistant professor, the second rung on the Harvard faculty ladder. Academic advancement at Harvard moves along a steep pyramid, and I was keen to reach the top. I continued my basic research on T cells and macrophages but now sought to balance my portfolio with experimental drug trials. My defeat in the gamma race taught me that you could labor for years in the lab and be blown away in a moment, especially by a powerful biotechnology company. There were no academic accolades for losing.

Competing in clinical research made the playing field more even. This was a match against other physicians in other medical centers, rather than legions of Ph.D.s in pharmaceutical companies. And with each clinical trial came a grant from the drug company that supported your salary and the salaries of the other participating physicians and nurses. Experimental drug trials seemed the best of both worlds. At the same time that needy patients like James Leahey could benefit from cutting-edge treatments, you were academically rewarded.

Our hospital was among the first to test genetically engineered gamma interferon in humans. I ran the trial in AIDS-related Kaposi's sarcoma and a colleague, Stephan Bougert, oversaw the study in melanoma. I was convinced Stephan and I would make medical history.

"I spoke to a nurse on the hospital hot line," James eagerly told me. "I'm perfect for the gamma test."

I told James that was our research nurse, Joann Karatosis, who coordinated the enrollment.

Most centers conducting clinical trials have such a hot line to inform patients of the available studies and to encourage practicing community physicians to make referrals. Patients are screened for eligibility with blood tests and a physical exam.

"You should get into the program," I said. "Your numbers fit the criteria."

James's face brightened.

"We're starting the next group in a week."

I then explained to James that there are three steps to the process. Phase I is the first assessment of the drug in humans. Although for patients like James these studies are their only chance for effective therapy, the phase I trial is designed only to identify the drug's side effects. The initial groups receive very low doses of the experimental agent: If these minimal doses appear safe, then subsequent groups are enrolled at increasingly higher doses. The aim is to define the maximum tolerated dose where intolerable toxic side effects occur. If you have the bad luck to enter early, it's likely you will receive doses too low to be effective. If you are enrolled late in the trial, you could very well suffer severe side effects that define the limit. James would, fortunately, enter in the middle group.

It is only after phase I that large numbers of patients are treated at safe doses. Phase II is where you determine clinical benefit. If the drug seems to work, and the side effects are tolerable, then phase III begins.

In phase III, the experimental drug is compared to standard therapy. If the experimental agent outperforms established drugs, then the FDA approves the novel treatment.

It is only after FDA approval that there is flexibility in usage. The approved drug can be prescribed at the discretion of the doctor, even for diseases that were not studied in phases I, II, or III. For example, if gamma interferon were approved for melanoma, as we anticipated, it could legally be given for breast or lung cancer if standard treatments were exhausted and the physician believed it prudent.

"It seems like a long haul," James grimly said.

I nodded in sober agreement. Many desperate people are turned away during the years it takes to finish the three phases. This was all the more painful in the case of gamma interferon. The drug was seemingly nontoxic in animals, and its effects in the test tube on immune cells were unparalleled. Still, it had to pass through three phases of clinical development. No drug company would dare deviate from this slow route and risk submitting incomplete data to the FDA; that would destroy their chances for approval.

James Leahey undressed, neatly folding his white shirt and paisley tie on the ledge next to the examining table.

"That shirt is one size larger than usual," he said as I began to palpate his neck.

James's lymph nodes were like eggs, distorting the curvature under his jaw. I carefully studied his retina with the ophthalmoscope, since melanoma sometimes spreads to the eye. There were no black deposits. When James breathed deeply, I heard a harsh whine, bronchi choked by melanoma. His liver was hard and enlarged, extending well below the border of his right ribs, and when he turned on his side, his cancer-filled spleen dropped against my fingers like a melon tethered on a string. The enlarged lymph nodes in his groin distorted the normal crevice and were growing into the hard muscle of his thighs.

I ended with the neurological examination. Although the CAT scan of James's brain two weeks earlier had been normal, melanoma can spread so quickly that it was possible a new deposit already had taken root. And the tumor can be insidious, causing minor changes in coordination, facial symmetry, or cognition. But, thankfully, I found nothing abnormal.

"Do I still qualify?" James asked tensely.

He did, I replied with broad smile. He should receive the first injection next week.

James, I told myself, was fortunate compared to others in his situation. Stephan Bougert and I had lobbied hard to get gamma. I knew the director of the pharmaceutical company developing the drug from past interferon conferences and presented a detailed package to him. It documented that we had a long track record in conducting phase I clinical trials, had the sophisticated laboratory expertise to measure the effects of gamma on the immune system, and could rapidly enroll patients from our clinics. Implicitly, a study conducted at our hospital carried with it the imprimatur of Harvard.

We were competing, of course, with other major cancer centers, each with prestige and resources. Only two sites would be selected, one on each coast. Stephan Bougert and I celebrated with Heinekens when we heard.

Stephan handled most of the melanoma referrals personally. It was a fluke that I saw James Leahey. I knew James's primary care doctor, and he had asked me to navigate James's entry into the study. I would put his name on Stephan Bougert's list personally.

Stephan was ten years older than I, a tall man with a square face, jutting jaw, and thick salt-and-pepper hair. His manner was distinctly acid, a style he claimed was inherited from his aristocratic French father.

"This fellow needs a mortician more than a doctor," Stephan said as he viewed the CAT scans of James's chest and abdomen. "Gamma can't resurrect the dead."

"Oh ye of little faith—I forgive thee for thou knowest not what thou do," I replied, only half jokingly.

"Cannon balls," Stephan continued, pointing to the large masses of melanoma. "In the lungs, in his liver, in his spleen." He paused and looked hard at me. "You're sure there's no spread to the brain?"

I put the CAT scan of James's head on the light box before us. The sculpted surfaces of the cerebrum and cerebellum were smooth and regular, and the deeper tissue of the brain uniform. Stephan was satisfied, then meticulously studied the blood test results.

As he did, I thought about Stephan's cynicism. I didn't believe it was entirely hereditary. Stephan had long conducted phase I and phase II trials, year after year enrolling desperate people into studies that were their only hope. And year after year he saw that hope crushed. Nearly all phase I and II trials failed. That was the nature of drug development, especially for cancer. Many experimental treatments, appearing acceptably safe in animal testing, proved terribly toxic in humans, even at low doses. And if unforeseen side effects did not kill the drug, then lack of efficacy did. On average, a successful new cancer therapy was found every decade or so. The anthracyclines, like Adriamycin, were developed in the early '70s, and despite significant cardiac toxicity, proved critical in the cure of leukemia and lymphoma. Cis-platinum was next developed, again a very toxic drug, poisoning the kidneys and causing deafness, but was the key to curing testicular and ovarian cancer.

I wasn't sure why Stephan persisted. Perhaps he believed that finally he would score—and was too guarded to admit that gamma interferon was the winner. Or perhaps he kept at it for the simple reason that this was now his job: something he did and did well, and at the end of the week was a paycheck.

"How's my man look for entry?" I asked.

"He'll be in the queue for the third dose group," Stephan replied. "There are four slots, and this week he's the twelfth eligible referral. We'll choose by lottery."

Lottery.

My stomach knotted. *Four out of twelve.*

"He's young and determined, a perfect study subject. Works as a claims clerk at an insurance company, incredibly organized and sincere. He'll keep all his appointments, report all his side effects, take every dose on schedule."

Stephan tightened his lips in silent consideration.

There still could be a lottery, but didn't we assign first priority to patients from within the medical center, the patients we personally cared for?

"All twelve referrals are from inside," Stephan said.

My stomach tightened further.

I would not ask Stephan to make an exception, as much as I wanted James treated. I thought back to other studies that were so deluged with referrals that we chose by lottery. An investment banker in his early thirties with AIDS told me plainly that he would give a million dollars to jump the queue. The money would be donated as a charitable gift to my laboratory or placed in a discretionary account outside of the hospital to use "in any way you like."

That was not the first such offer of a large sum, only the most bald-faced. I declined politely, as I wondered what strings I would pull to try to save my own life were I in a similar position.

Stephan, at the weekly meeting of the clinical research team, had agreed with me about the banker. No one countenanced personal gain, but several other members on the team argued that medical centers also needed patrons—like colleges admitting students from wealthy families whose gifts are translated into scholarships or new buildings. The money given to our program would support more research and ultimately serve a greater good. But these arguments did not prevail. No one was naïve about the importance of philanthropy in medical science, but it could not place one life above another.

The other temptation was not money but love—the real affection a physician feels for certain patients. But the rules were ironclad: slots could not be assigned based on personal favorites.

Snow was falling the morning I sat with James to obtain his informed consent for the study. His face had a crimson glow from the cold. The seven-page document, written in lay language, outlined the rationale and conduct of the phase I study.

James winced as he read the first paragraph.

> I understand that I have advanced cancer for which there is no cure and for which there is no satisfactory alternative therapy. I understand that I am being considered for one of the first clinical trials of human gamma interferon. The effects of this therapy, although studied in animals, cannot be predicted in humans, and may vary from person to person. I understand that the primary aim of this study is to assess the safety of the treatment, and that I may have no benefit from it.

James stopped.

"I know all this."

I shook my head and said that he had to read it, in full. This was required by law.

He returned to the page, his finger following the lines. "It says here there is no commitment to continue to provide me with the treatment beyond this phase I trial, even if I improve on it."

I nodded gravely. Although every effort was made to provide an experimental drug to patients who were benefiting, the therapy was entirely in the hands of the pharmaceutical company. Clinical investigators like myself had no freedom in terms of its administration. If drug supplies were scant, or the company decided to target different cancers than his, they could cut off the treatment after the phase I study was over. They didn't usually do this—it was bad PR. But it could happen.

"Your group will be chosen by lottery."

"Lottery?" James blurted.

"Yes. A lottery."

His face clouded.

I spoke in an even voice, trying to mask my anxiety. I explained that usually we entered our own patients on a first come, first served basis and then opened the trial to outside referrals. But gamma was so long awaited that the phase I trial was oversubscribed with internal referrals. A lottery was the only fair way to choose.

"You can't do anything about it? You're the chief of oncology."

I said I couldn't, despite my position.

The drawing was set for Friday morning around eleven o'clock. I repeatedly glanced at my wall clock while I occupied myself with paperwork. Around ten to eleven, my secretary, Youngsun, entered with James behind her. He was dressed for work, with gray trousers, white shirt, paisley tie, and a blue sports coat.

I put my paperwork aside and invited James to sit down. He took the chair, his head bowed.

"I just couldn't wait at work," he said.

I nodded and said the call should come shortly. I asked how his mother was holding up. James said she was anxious but hopeful. I knew he was her only child and that his father had died many years before.

"It's Joann from the clinic on the phone," Youngsun announced on the intercom.

"Transfer the call."

I picked up the phone on the first ring.

"Mr. Leahey wasn't chosen," Joann said heavily. "I'm sorry."

My eyes closed at the news. I quickly opened them and shook my head.

James's face twisted in anguish, and his hands began to tremble.

I moved to him. He stood up unsteadily. I placed my hands on James's broad shoulders. He stepped back, his face turned away.

"I'm disappointed, too," I began, my words sounding thin and useless.

"For now, James . . . for now, we have to buy time." I paused. "Often phase I studies are expanded. More patients are enrolled. You're number six in the queue. Four will enter. It's possible in the next few weeks two more slots will open.

"Some people in front of you may need to drop out of the study. They'll have to be replaced. This is not unusual since"—I searched for the gentlest phrase—"since their disease often spreads so quickly, they'll need radiation and steroids. That disqualifies them from staying in the trial."

James Leahey sat mute, his eyes wide and frozen.

"We have three options here," I said, my voice becoming firmer. "The first . . . the first is to do nothing. The one-way ticket to Hawaii."

I began with this, to shock him into paying attention. James's eyes widened with deeper alarm.

"The second is chemotherapy. It has significant side effects and rarely shrinks the tumor for a significant period of time."

James shook his head, indicating this was not what he wanted.

"And the third is alpha interferon. We know gamma looks far superior, but there are documented cases of regression of melanoma with alpha. Some of the responses last for many months, even a year."

I paused to give him time to focus on that chance.

"I think that's the best way to go. We'll try to hold the cancer at bay with alpha until we can get gamma."

James's lips trembled. For a long moment he was unable to form a reply.

"I'm going to lose my life . . . because a nurse a picked a different slip of paper!" Tears welled in his eyes. "God dammit to hell, Doc, it's not fair. It's just not fair."

His face was lit with anger and his chest heaved.

I tried to hold his fierce gaze with mine but felt weak and wavered. I reached again for his shoulders. For a long time I said nothing, just pressing my grip. His breathing gradually slowed. I removed my hands and reached for the box of tissues on my desk.

"I'm sorry," James said, blotting his eyes.

"You're not the one to say 'sorry.'"

Five days later, James and I walked together from my office along Longwood Avenue to the oncology clinic. Thick clouds filled the slate March sky. James lifted the collar of his overcoat to shield his neck from winter's icy tongue. We navigated the sidewalk cautiously, stepping over sooty mounds of snow left from the prior week's storm.

As we neared the entrance to the clinic, I took James's arm just above the elbow. The receptionist said we were expected in the treatment area.

Helen Gallagher, a tall woman with jet-black hair and a commanding manner, greeted us with a welcoming smile. She was one of our most senior oncology nurses.

"Irish?" James asked.

"Of course," Helen replied, quickly studying James's face. "I also married one."

"Maybe you'll bring me more luck than I've been having."

Helen's eyes briefly met mine.

"I believe in luck," she replied evenly, "and in doing everything possible."

She led James to the far corner of the treatment area. I followed at his side. It was a typical busy Thursday in the clinic. Treatments are often given on that day, so the patient can recuperate over the weekend with family around to help. I watched James's face tighten as we passed more than a dozen people with cancer.

This first trip through was always jarring. Men and women, young and old, sat in blue vinyl reclining chairs, forearms extended on boardlike attachments. Stark metal poles stood next to the chairs, holding bags with yellow and red and clear solutions of chemotherapy that percolated down through plastic tubes into the braced forearm veins. The overhead fluorescent lighting shone off the many bald heads we passed. A nurse attended to each patient, regulating the infusion, checking blood pressure and temperature, inspecting the vein on the forearm to be sure the site was secure.

"We'll teach you to inject yourself," Helen said. "Alpha interferon is given like insulin, beneath the skin, not into a vein."

She had arranged the necessary equipment for James's instruction: syringes with thin needles, a medication vial with a closed lid, alcohol swabs, and a pliant rubber ball. James would first practice drawing up the right amount of solution from the vial into the syringe, change needles, and then stab the ball at the correct angle and depth. When he was adept at this, Helen would select a site on each thigh and have him inject himself with saline. He would take a similar volume of alpha interferon five times a week, just before bed and with Tylenol.

"Here's a prescription for the alpha," I said to Helen. "And another for the needles and syringes."

She placed them in the pocket of her white coat. Helen would contact James's pharmacy after the teaching session. The drug needed to be ordered: No pharmacist routinely stocked such an expensive and infrequently prescribed medication. A regional distributor would courier it for overnight delivery.

I said good-bye to James. His expression was fearful, like a child's returned to bed after a nightmare.

Shortly after nine the next morning my phone rang.

"Jesus, Mary, and Joseph," James exclaimed. "I went to the pharmacy just after it opened, at eight. They have the alpha from the distributor, but the drug requires preapproval by Massachusetts Health and Life. It was denied. I spent the better part of an hour working my way up the chain of command. I'm their employee!"

James finally reached the chief medical officer of the plan, Dr. Philip Henderson. It was a brief conversation. Alpha interferon, Dr. Henderson stated, was approved by the FDA only for Kaposi's sarcoma and hairy cell leukemia, not for melanoma. The insurance company would not cover such expensive treatments outside their licensed indications. My prescription fell outside their guidelines as not "medically appropriate."

I was stunned. HMOs had taken a severe look at the therapies that they would pay for. But James wasn't in a managed care plan. He had private insurance, the equivalent of my Blue Cross/Blue Shield. The bean counters there must be tightening the rules, too.

"I need the mailing address of the chief medical officer, Dr. Henderson, and the names of your company's president and CEO. I'll fight full force with you. But I also need to make a few personal phone calls as backup."

26 Harbor Plaza
Boston, MA 02110
ATTN: Dr. Philip Henderson
cc: Mr. Paul Paige, CEO
        Mr. Carl Long, President

Dear Dr. Henderson:

Mr. James Leahey is a patient under my care who has metastatic
melanoma. I believe at this time the best therapy for Mr. Leahey is
treatment with interferon alpha. This recommendation is based on the
efficacy of treatment of malignant melanoma with interferon alpha, the
risks associated with chemotherapy, and the lack of other meaningful
options.

As Chief of the Medical Oncology Program here, and as faculty at
the Harvard Medical School, it is my considered opinion that
chemotherapy for malignant melanoma with metastases in lymph
nodes, lung, liver and spleen would cause significant toxicity and offer
little likelihood of benefit. The oncology literature has a number of reports of clinical trials utilizing alpha interferon in the therapy of malignant melanoma. Indeed, studies in the United States and Europe
documented response rates of 15–20% (see attached reference 1 from
Cancer Research). Alpha interferon is generally well tolerated in patients with melanoma, and responses appear to be best among patients
with Mr. Leahey's particular situation, that of soft tissue and pulmonary
metastases (see attached reference 2, abstract from the recent *Proceed-
ings of American Society of Clinical Oncology*).

The optimal treatment for Mr. Leahey is 18 million units of interferon alpha A which costs approximately $150. A weekly treatment
consisting of 5 days of subcutaneous doses would cost $750. I would
point out that the cost of chemotherapy given for his malignant
melanoma would be approximately the same amount.

Mr. Leahey's prognosis is very poor without interferon alpha therapy.
Although the chance of complete regression of his cancer is small,

there is precedent in the literature for the application of this treatment. Interferon alpha, while not approved specifically by the FDA for melanoma, is widely used in medical practice for this diagnosis.

I am including a document from the Health Care Finance Administration (HCFA) which clearly states, based on a recent ruling, that there is no question about the legality of utilizing interferon alpha in situations outside of FDA indications, such as Mr. Leahey's. I believe that there is a responsibility to this patient to reimburse him for what is his best medical therapy. Failure to meet that responsibility would indicate a breach of trust and the spirit of his insurance coverage.

I look forward to discussing his case in greater detail with you to assist in the evaluation of this claim.

Sincerely,
Jerome E. Groopman, M.D.
Chief, Division of Hematology/Oncology
New England Deaconess Hospital
Assistant Professor of Medicine
Harvard Medical School
cc: Mr. James Leahey

The letter was written to push as many buttons as possible, to simultaneously impress, document, cajole, and threaten. James had little hope the letter would work. He didn't want to wait for a response. He was ready to publicize his plight, write letters to the *Boston Herald* and the *Globe*, try to get on local radio talk shows, even picket in front of the main headquarters of Massachusetts Health and Life.

"Easier to catch flies with honey than vinegar," I said, quoting my grandmother's favorite adage, but uncertain if I could sell alpha's allure.

___⌒⌒◞

"Let's speak candidly here, Dr. Groopman," Dr. Philip Henderson said.

I mailed him my letter certified, and the day the return receipt arrived back in my office, I called. When I did not hear back after two hours, I placed a second call, pointedly telling Dr. Henderson's secretary that an urgent reply was needed. I told her that I expected to hear back that day. Dr. Henderson should page me if for some reason I was not at my desk. I would also leave my beeper on after work hours waiting for his call.

James inquired about Dr. Henderson through other workers at the company and I found his biography in the listings of the local medical organization. Like many in this role, Philip Henderson was an older man, retired from active clinical care. He had been a general surgeon practicing in a small town north of Boston. There was nothing to suggest he was expert in cancer, or experimental therapies. James said Dr. Henderson was known as friendly to his staff and once helped a secretary get a referral to a top neurosurgeon when her husband had an aneurysm.

"I'm sure your heart goes out to this fellow," Dr. Henderson said. "But the company has started to deny paying for unlicensed treatments that aren't likely to be significant. There's nothing I can do. Those are my marching orders."

"As I documented in my letter," I began in a scripted reply, "there can be regression of metastatic melanoma with alpha interferon. The success rate is low, Dr. Henderson, but it's real. I personally led several of the early clinical trials. A lovely middle-aged woman, a schoolteacher from New Hampshire, with extensive disease like Mr. Leahey's, involving lymph nodes, lung, and liver, had regression of the cancer that lasted nearly a year. It meant a great deal to her and her family to have those extra months."

"I'm sure it did," Dr. Henderson said sympathetically. "But I'm not in a position to make policy. Anecdotes fall on deaf ears here. The higher-ups want data from clinical trials and want to pay only for FDA-approved drugs."

I paused to consider my reply. The "higher-ups" ignored a dimension of treatment that could not be distilled from the outcomes of clinical trials or FDA rulings. James could not peacefully face the end if he were filled with regret that he didn't fight with whatever weapons were available.

This argument, though, would "fall on deaf ears," and I decided not to pursue it.

"My patient has extensive disease," I continued. "If alpha interferon is going to work, we'll know that in two to three months. I agree it's not a potent therapy, but it's the best there is for him. Two to three months won't cost the company that much."

My words were met with a brief silence.

"Dr. Groopman, I was in practice for forty-two years," Philip Henderson said in a gentle tone. "I realize you have to give the patient and his family something to hang on to. Consider some light chemotherapy for this man. Not a dose to cause side effects. Essentially a placebo. That's really what you're saying about interferon. Mr. Leahey will have hope, and it'll take the pressure off you."

I saw older doctors take this tack during my training as a medical student and intern. That past generation, raised at a time when there was little genuine therapy to offer, believed their actions to be merciful. It was linked to not telling patients and their families the true prognosis. For a while the agony was defused. But the lie robbed the patient of choice in his treatment, and in how he spent his remaining time. When the truth became apparent, he knew he had been betrayed. His doctors could never again be trusted.

"That's not my style."

I stopped from saying something that would alienate Dr. Henderson.

"James Leahey has been a longtime employee of your company. I thought there would be some discretion. It wouldn't look good if this came to light, that Massachusetts Health and Life appeared uncaring after all Mr. Leahey's years of dedicated service."

"You don't know the corporate mentality," Dr. Henderson

replied, his voice still collegial. "Mr. Leahey can find himself a lawyer or speak to the papers. It would stir things up for a few days. But the president would ask me about the course of melanoma and realize Mr. Leahey won't be around, and it'll all blow over. I'm sorry, Dr. Groopman."

I replayed our conversation and assumed the role of devil's advocate. Philip Henderson was now in a corporate position. He was constrained to follow the orders set by his superiors. Health care was a business. Clinical decisions had to be business decisions. *Just wait until the CEO of Massachusetts Health and Life is sick,* I said to myself. Then we would see if the devil were given his due.

I concluded we would get nowhere with the company. James's disease was progressing quickly. If alpha had any chance of helping, it needed to be started immediately. My ace in the hole was the pharmaceutical companies that sold alpha interferon. Each had a program to donate the drug to indigent patients. These companies already had made many millions of dollars from the agent, and projections for the future market were close to a billion. Such profits would come largely from therapy for infectious diseases, like hepatitis, not cancer, since alpha appeared to be potent against certain viruses. The giveaway program generated goodwill around a costly drug.

I knew the physicians in the companies producing alpha interferon and hoped a personal plea on James's behalf would secure the drug for him. Their higher-ups would calculate that with melanoma in the liver, lungs, and spleen, the company would only have to provide a few weeks or months of the drug gratis. It was an imperceptible blip on a vast balance sheet, and a small favor to a Harvard hospital that had helped develop their drug.

I flipped through my Rolodex and dialed Claire Vitale in New Jersey. Her company was one of three in the United States that produced alpha interferon, and we had worked closely together on the Kaposi's sarcoma studies. Claire had been an oncologist at a university hospital and left for a senior position in industry. She had broad responsibility not only for clinical trials but also for business-related matters, such as physician-targeted marketing and regulatory issues with the FDA. Her diminutive stature and soft voice belied an iron will. Despite wearing several corporate hats, Claire had never relinquished her clinician's sense of the primacy of the individual patient.

We exchanged the usual pleasantries, asking after each other's children and spouses.

"I need a favor," I said, knowing that Claire liked to hear things fast and straight. "I have a man in his thirties with widespread melanoma. He's a claims clerk at an insurance company here, and I doubt he makes more than $30,000 a year. A lovely guy, he has no illusions about his situation. He wasn't chosen for our gamma study."

"That's too bad," Claire answered. "And now he's having trouble accessing alpha, right?"

"Exactly. I wrote a detailed letter to the insurance company, and just talked to the chief medical officer. They won't budge."

"The reimbursement issue is going to heat up around all the new biotech products. And not only at HMOs. I'm setting up a group here to do nothing but deal with that."

It was a clever strategic move on Claire's part. Genetically engineered products were expensive, because proteins are much more costly to produce than traditional chemical drugs. Alpha was the first, and how it was reimbursed for non-FDA indications would set a precedent in the marketplace and drive corporate revenues.

Claire explained that her "working group" would provide detailed documentation to an insurance company to justify requests like mine—so-called pharmacoeconomic data. This would include calculations balancing the cost of treatment with alpha against the costs of chemotherapy or radiation, and the relative impact of each on quality of life and productivity. Claire confirmed what I put in my letter: the costs of standard chemotherapy, which no insurer would deny, were comparable to alpha treatment, because with chemotherapy the patient had to come into clinic regularly to be treated and required more ancillary therapies, like antinausea and sedative medications. With alpha, the patient self-administered the drug at home, and aside from the nightly Tylenol for fevers and chills, required little else.

"This is the best way to get their attention. Hit them with the bottom line. Now, tell me more about your patient."

I summarized the case, leaving out no detail. I emphasized that James was a proud man and a fighter, that I'd offered other options to him, that we acknowledged the best scenario was to buy enough time until he might enter the gamma trial.

"Like I said, this group to handle reimbursement is not fully formed. But I'll put one of my people to work on this immediately. Meanwhile, you know we have the giveaway program. But he's making $30,000 a year, and he's not really there."

"That's why I said I need a favor, Claire."

We reviewed the games that were played by doctors and patients with regard to payments. James could transfer his savings to his mother to appear indigent. Or he could pay for alpha out of pocket until he was bankrupted. This was a slow and degrading process. And, I knew, James was supporting his mother.

"I need a letter from you," Claire said. "It has to clearly state that in your clinical judgment alpha interferon is the best option for this man, and that given his income, and his family obligations, it's a financial hardship for him to pay for the drug. Make it pull at the heartstrings. I'll pass it up the chain of command and try to work something out. You have my word, Mr. Leahey will get the drug ASAP."

I thanked Claire warmly. We then turned to working out logistics. The medication would be shipped directly to the hospital's general pharmacy. It then would go to Helen Gallagher, who would pass it on to James. There was considerable paperwork required. I would have to keep records of James's doses, side effects, and the course of his cancer.

"His insurance company has to pay for blood tests, CAT scans—all the usual clinical monitoring for treatment of melanoma," Claire said.

I said Massachusetts Health and Life would. So-called standard clinical care was routinely reimbursed.

"You're an angel, Claire," I said as the conversation ended.

"Don't let that out. To get promoted here, they have to think you're a hard-boiled bitch."

The first weeks of therapy were difficult for James. The high doses of alpha interferon, although administered in the evening and with Tylenol, caused worse side effects than expected. He awoke in the middle of the night drenched in sweat, his temperature above 101, his muscles in painful knots. James tried to work during the first weeks, but lack of sleep and persistent fever and muscle aches made it impossible. He finally took a medical leave.

His mother, Mrs. Leahey, who lived in the apartment above his, attended to him. She accompanied her son to each clinic visit. She was a large woman with silver gray hair, searching brown eyes, and a forceful manner. Her immediate concern was to make sure James was getting enough fluids and nutrition. His appetite was poor due to the treatment.

"I never thought she'd be able to inject me," James said, eyeing his mother.

When James's hand had become too unsteady, Helen Gallagher had taught Mrs. Leahey how to sterilize the skin and insert the needle, delivering the proper dose of drug each time.

"Now, did I raise a son to call his mother *she?*" Mrs. Leahey chided in her lilting accent.

"You sound like my mother," I said.

"Doctor, we all went to the same school."

I examined James, focusing on the neurological assessment. I didn't want to assume his unsteady hand and weakness were due to the treatment. They could be the first signs of melanoma in his brain or spinal cord. But there was nothing from the tests to suggest that.

I encouraged James to hold on, saying that the first few weeks are always the worst. There is a phenomenon called tachyphylaxis, a clinical term meaning that with many toxic drugs the body adapts with continued treatment. We hoped the side effects would lessen.

"Like I told you, we Irish are tough," James said in a wan voice. He could put up with the suffering, so long as there was hope.

I replied there still was.

We set a schedule of weekly visits. I did this both to keep a close eye on James's clinical status and to provide psychological support. Mrs. Leahey did not drive, so they had to find a neighbor to bring them in.

At his three-week appointment, James still moved slowly and his eyes were dim. There was little tachyphylaxis: the fevers and muscle aches only fell to an 8 on a scale of 10, he calculated. The rock-hard lymph nodes had not budged. His lungs rattled harshly. His engorged spleen again arrogantly struck my hand when he turned on his side.

"You're sure you can put up with the side effects?" I asked. The unspoken question being that perhaps it was all in vain, that the time remaining was being squandered.

He looked at me uncertainly.

"I can."

James missed two visits in a row, the first because of a ride falling through, the second because I was called out of town unexpectedly.

It was early April when he was set to return. Heavy sleet had been falling since the early morning. I called him at home and said we could reschedule for the next week. But he was insistent. I offered to send a wheelchair to meet him at the hospital entrance, but James wanted to walk with his mother from the parking garage despite the weather. He stated the side effects were lessening, particularly the muscle aches, and the height of the fever was now only about 100 at night. The exercise would do him good.

"I'm breathing easier at night," James said as he lowered himself onto the examining table. Mrs. Leahey confirmed that was true. She didn't hear as much coughing. And he was eating better.

I assumed there was finally tachyphylaxis. Without fever, his lungs were less stressed and his appetite improved. But as I ran the tips of my fingers over his neck, first gently and then more firmly, pressing between the strap muscles below the jaw, I was stunned. I quickly palpated the wells above his collarbones. I moved my left hand into his left armpit and then reached over with my right hand and dug deeply into the right.

"My God, they're nearly gone!"

"I thought so, too," James said with a wary smile. "But you read all this stuff about being sick and believing things that aren't really there."

Mrs. Leahey began to cry.

I excitedly resumed the examination. His lungs retained only a faint whisper of their prior cacophony. His liver and spleen were pliant and had retreated to their proper places under his ribs.

"In just the last three weeks," I marveled.

"Mom almost burned the church down with votive candles," James said with a smile.

Mrs. Leahey's moist eyes met mine. I nodded my head in reverent affirmation.

I rapidly detailed the next steps. We needed objective confirmation. A chest X ray should show disappearance of the large masses that had been in his lungs. A CAT scan would document the regression of the cancer in his liver and spleen. We would also repeat a scan of the head to assure that no disease was lurking there. With those data, we would be armed for battle with Dr. Henderson and Massachusetts Health and Life Insurance.

* * *

"You received my package?"

"Indeed, I did," Dr. Henderson replied.

I waited for him to continue.

"Remarkable. A medical miracle."

It was.

"I'll sign off on it today," Dr. Henderson said. "According to his policy, he has another nine months of medical benefits. If he's still on leave, then his coverage expires. Other arrangements will have to be made."

* * *

It did not take that long. After two more months of therapy, I witnessed something I have never seen again: complete regression of widespread melanoma. Every deposit cleared, in lymph nodes, lungs, liver, and spleen.

James had adapted to the injections, free now of fever and muscle aches. He was still fatigued, and took afternoon naps to recoup his energy.

"How long do you think the remission will last?" he asked.

I honestly did not know. The very best responses reported in the medical literature were with less extensive disease, and lasted for up to a year.

"In the meantime," James said, his gaze unwavering, "I'm here."

I retrieved his original biopsy slides from the Pathology Department. There was no doubt about the diagnosis. Multiple sections of the lymph nodes had been made, all showing the ugly ink-black cancer cells obliterating the normal tissue. I asked the attending pathologist to make new cuts from the blocks of preserved tissue and study the cells for any unique characteristics. But James's melanoma was found to have the same repertoire of proteins, sugars, and lipids that mark the cancer in its typically unresponsive form. There was no clue from the tissue to explain the miracle.

Nine months approached, and James's benefits from Massachusetts Health and Life were due to expire. The social worker in the oncology clinic helped him apply for Social Security Insurance. I was determined not to interrupt his treatment and sensed there would be no grace period with Massachusetts Health and Life. James knew a claims clerk at the federal office. He approved James quickly for long-term disability, and the cost of the drug was reimbursed by the government.

It was a crisp bright autumn day in 1998, and I had to shield my eyes from the sun as I scanned my patient roster. James Leahey's name headed the list. It was now thirteen years since we met. His rich black hair had become thin and laced with strands of gray and a pair of reading glasses rested in his breast pocket.

James had received a full ten years of alpha interferon, and then, with mutual trepidation, we decided to stop. I checked him every two months for a year after ceasing therapy, each time fearing my fingers would find a growing lymph node, or my ears would hear the harsh gasp of a choking bronchus. But the year passed uneventfully, as did each subsequent year. His miracle endured. The expected miracle of gamma interferon never came to pass.

The trials in human cancer were a dismal failure. Despite gamma triggering the patients' T cells and macrophages, not a single significant and lasting regression of a tumor was observed. I at first insisted some technical aspect of the nationwide trial must be flawed—the preparation of the genetically engineered protein was wrong, or the dose or schedule of administration misconceived. But that was not the reason. Gamma had no meaningful effect on human cancers. Men simply were not mice.

"Pygmalion," Stephan Bougert said. "You were deeply in love with your own work."

I admitted he was right. I was intoxicated by the laboratory science when I should have scrutinized the research with humility.

I continued doing clinical trials and over the decade developed a more measured sense of hope. Several genetically engineered drugs proved beneficial in treating cancer and AIDS.

James was working part-time in an accounting firm while taking courses in business administration at Northeastern University. At follow-up visits, after inquiring about my ongoing research and outlining the progress toward his career goals, he would turn to his main concern, his weight. He had developed a middle-age spread, and despite daily morning exercises, found it hard to trim down. I empathized with him, noting that I had begun a similar battle, with little success.

"I still don't know what to make of it," James said at that autumn visit in 1998 when I told him I wanted to write his story. "It hasn't made me more religious, like it has Mother. She's become almost mystical."

Mrs. Leahey believed that a guardian angel steered the research nurse's hand away from her son's name in the lottery and led him to the therapy that saved his life.

"I do believe God was good to me," James went on. "But I don't see why He should've been better to me than to other poor souls who didn't make it."

"I've told your story to dozens of senior physicians," I said.

Each was initially incredulous, I informed him. Then they recounted a similar case, over many years of practice, of a patient with an incurable cancer who, against all odds, survived.

James and I sat together silently for a long moment, drinking in the mystery. Finally, he spoke.

"It's a pity you never learned why. Then you could have shared the miracle."

Don't Just Do Something—
Stand There!

⁓

Alex Orkin was told he had less than six months to live unless he underwent a bone marrow transplant. A thirty-eight-year-old physicist at a university in New York, he came to me for a second opinion on a frigid February day in 1994. He was a short, stocky man, with receding black hair, a broad nose, and darting eyes behind thick glasses. Some eight weeks earlier, at a faculty Christmas party, he had felt dizzy and nearly fainted. He had been fatigued for several months but assumed it was the long hours at work and the stresses of parenting three-year-old twin boys. The cause of his symptoms, though, proved to be his blood. He was very anemic, with less than half the normal number of red cells. His white blood cell and platelet counts were also low.

Reduced numbers of circulating blood cells can occur by two general mechanisms. One is impaired production of blood in the bone marrow. The other is loss or destruction of blood outside the marrow. Dr. Frank Hochman, a senior hematologist at a medical center in Manhattan, had performed an extensive evaluation. I knew Frank from scientific meetings. He had a reputation as an intense, hard-driving academic clinician and led the marrow transplant program at his hospital. Frank had found no indication of blood loss or destruction. Rather, the biopsy of Alex Orkin's marrow showed it was largely depleted of cells and extensively scarred. This suggested the marrow was incapable of producing sufficient blood.

"Marrow failure state—myelodysplasia? aplastic anemia?" Frank Hochman had written in his notes.

Frank Hochman was uncertain about what was causing the marrow to fail. Myelodysplasia is a frequent cause of aborted blood cell production. The marrow is somehow injured, so the remaining cells die prematurely. The injured cells appear distorted in size and shape (*dysplasia* means "distorted form" in Latin, and *myelo* refers to marrow). Myelodysplasia can be caused by toxic chemicals, drugs, or radiation, damaging the cells' DNA and thereby deforming their appearance. Many cases of myelodysplasia evolve into acute leukemia.

Aplastic anemia is a disorder where the marrow appears empty, with very few cells (*aplastic* in Latin means "without form"). Survival is short, because so little blood is made. Patients succumb to infection from lack of white cells, or to hemorrhage from lack of platelets. Aplastic anemia can be caused by the same injurious factors that cause myelodysplasia: chemicals, drugs, and radiation, as well as by certain destructive viruses and autoimmune disease, like lupus.

The question marks in Frank Hochman's notes reflected the difficulty in making a diagnosis. The sparse numbers of marrow cells, Frank wrote, could signal aplastic anemia, while the changes in their appearance suggested myelodysplasia. There were too few cells to analyze for DNA damage, which would support the latter diagnosis.

Alex Orkin received transfusions of red blood cells to ameliorate his anemia and transfusions of platelets to prevent bleeding. An empiric course of treatment with immunosuppressive drugs, which is sometimes effective in aplastic anemia, had no discernible benefit.

For a person less than forty years of age, bone marrow transplant is the best curative option for marrow failure due to either myelodysplasia or aplastic anemia. But Alex Orkin had no related donors: he was an only child, and no extended family members genetically matched. Furthermore, the national registry of marrow donors, despite a roster of three million volunteers, had no one who was compatible. So Dr. Hochman recommended that Alex Orkin undergo an unmatched transplant, a procedure with very high risk. First, as in a standard transplant, all of Alex Orkin's marrow cells would be destroyed by high doses of chemotherapy and radiation. Then the donor's stem cells would be infused. These donor cells home into the emptied marrow cavity and repopulate the entire blood, growing into red cells, white cells, and platelets. But as the donor T cells grew, they would perceive the host as very foreign because of the genetic incompatibility and begin to viciously attack host tissues. This is called acute graft-versus-host disease. Alex Orkin, as the host, would have his liver and bowel and skin ravaged by the grafted T cells, and he would develop hepatitis, colitis, and severe dermatitis. Powerful drugs would be given to try to temper the attacking T cells, but many patients die from acute graft-versus-host disease, and those who survive usually have a chronic debilitating syndrome.

Heavy sheets of sleet began to fall as Alex Orkin and I reviewed his medical history. His voice was somber, and when he stated the dire six-month prognosis Frank Hochman had given him, he clasped his hands tightly to arrest a growing tremor.

Alex Orkin was born in Queens, New York, to a middle-class family—his father an accountant, his mother a schoolteacher. Both parents were in good health and there was no history of blood diseases. He had been an undergraduate and graduate student at MIT. His work was in the theoretical physics of subatomic particles. "I work entirely in my head, not in a lab," he explained. He knew of no possible radioactive or chemical exposures that could have injured his marrow: he did not garden or handle pesticides, never visited a nuclear reactor site, did not paint or varnish wood with volatile solvents, and took no regular medications. Since certain viruses can cause marrow failure, I asked about hepatitis or mononucleosis, but he never had these infections; nor was he recently vaccinated or suffering from a viral syndrome before his symptoms at the holiday party. His only travel was to Europe and Japan for scientific meetings, not to more exotic parts of the globe.

"But I was never in southern Japan, in the prefecture of Kyushu, or in the Caribbean," he offered.

I noted this in my record. He had anticipated my question about Japan and the Caribbean region endemic for a recently discovered virus, HTLV, associated with certain types of leukemia.

"And I don't use illicit drugs," he continued. "I smoked some marijuana at MIT but didn't like it much. I have no tolerance for alcohol. And my sex life has been sedate for someone of my generation. I never had a venereal disease. I am not homosexual. My wife, Sandra, and I met in college—she's an electrical engineer. I've been monogamous ever since."

Occasionally, I said, so-called autoimmune diseases, like lupus, can suppress the marrow.

Alex Orkin elaborated that he had no symptoms or signs of lupus-like diseases, specifically no arthritis, urinary problems, skin rashes, or visual difficulties.

He had some lower-back strain from lifting his children.

"You didn't take medications like Butazolidin for the strain, did you?"

This anti-inflammatory painkiller, prescribed for joint problems and lower-back strain, was associated with aplastic anemia.

"No, just Tylenol," he replied.

I asked if he were certain.

Alex Orkin raised his brow quizzically as his eyes bore into me.

"Of course I'm certain. I have a near-photographic memory. I forget no fact, regardless of how trivial."

I was disturbed rather than heartened by this. His medical history had been recounted too fluidly, and each of my questions had been anticipated. Alex Orkin was reading from a mental script, composed from Dr. Hochman's consultation, while I was seeking a fresh recounting, from true memory, hoping to find new information to clarify the diagnosis.

"I've prepared a summary sheet for you," he said.

It was computer generated. At the top of the page was "December 17, 1993," the date of the faculty party. Dr. Hochman's office address and telephone and fax numbers followed. Then there was a graph with three different-colored lines, one for red blood cells, one for white blood cells, and the third for platelets. The graph's vertical axis was marked "Numbers of Cells." The horizontal axis was labeled "Time," and was measured in weeks since the Christmas party. Arrows below the horizontal axis indicated important events: blood transfusions, bone marrow biopsy, genetic typing to search for a compatible marrow donor, beginning and end of the trial of immune suppressive treatment.

"May I keep it for my record?" I asked.

"Sure. I have it on a disk."

Alex Orkin's physical examination, aside from the expected pallor of an anemic man, was normal. There were no enlarged lymph nodes or spleen, no rashes or other skin lesions, no growths in his mouth or pharynx, and no masses in his testes or rectum—thus, no signs of an infectious, autoimmune, or malignant disease.

While he dressed, I took the slides of the bone marrow biopsy done in New York to the microscope room next to the clinic.

Marrow exists in the cavities of the large bones and is composed primarily of two elements enmeshed in a reticular matrix: maturing blood cells and nourishing fat. It is usually sampled by inserting a large trocar through the bone of the pelvis, and two preparations are made. The first is called the aspirate. This is the juice of the marrow, sucked up, or aspirated, from the cavity through a syringe attached to the trocar. The aspirate shows the diversity of developing blood cells, from the youngest, called blasts, to the fully mature that are released into the circulation. The marrow cells in the aspirate can be analyzed for abnormalities in their DNA, the signature finding in myelodysplasia. The second marrow preparation is a biopsy. Here, the trocar is used to bore out a core of the inner matrix. The biopsy reveals the structure of the marrow, how different cells are located with respect to each other, and what proportion of the marrow is composed of maturing blood cells versus nourishing fat.

Dr. Hochman obtained only a biopsy; he was unable to aspirate Alex Orkin's marrow. This is called a dry tap. It signifies that either there are too few marrow cells in the cavity to retrieve, or that the cells are trapped by scar tissue within the matrix and cannot be sucked up by the syringe.

I began the examination of the marrow biopsy under low magnification. Instead of a rich field of growing cells, it looked like a desert landscape. Large desolate tracts extended from the rough borders of the surface bone. Several wide bridges of scar tissue crossed these tracts.

I switched the microscope's lens to higher power to search for signs of life. Under this greater magnification, I slowly marched down the empty tracts. I finally found a few groups of immature marrow cells, clustered along deposits of fat, like bands of desperate nomads clinging to drying oases.

The picture didn't quite fit, and I wasn't sure of the diagnosis after examining the biopsy. So I took the slides to Ned Waterman, our pathologist expert in hematological diseases.

Ned Waterman carefully studied the biopsy, shifting from low to high magnification as I had done, and then moved his eyes away from the microscope.

"It's not myelodysplasia and it's not aplastic anemia," he asserted.

Ned had confirmed my sense that the picture was not clear.

Ned tightened his brow and his gaze drifted away from mine. After a long, pensive silence, he looked back at me and found the words to deconstruct his impression. Bridges of scar were often seen in myelodysplasia. But, Ned believed, they were a red herring here, since the few remaining marrow cells lacked the characteristic distortions of myelodysplasia. The cells looked a bit frayed around their edges but not intrinsically deformed in their nuclei.

"And why don't I think it's aplastic anemia?" he asked rhetorically. "Too many nests of marrow cells around the fat deposits."

Ned paused gravely. "No, it's my bet that something *outside* the marrow is suppressing this man's capacity to produce blood. Intrinsically, I believe, his marrow cells are healthy."

"Were flow studies or DNA analysis done?" Ned asked. Flow studies are special tests done to identify aberrant cells, such as those in myelodysplasia. DNA analysis involves examining the cells for breaks in their chromosomes.

"They were indeterminate," I replied. There had been too few cells to obtain a satisfactory analysis.

Ned and I combed through the remaining reports from New York. The tests for lupus and other autoimmune disorders were negative. Viruses known to suppress the marrow—HIV, HTLV, hepatitis, Epstein-Barr, cytomegalovirus, and parvovirus—were all sought, and none was found.

"Let's be doubly sure we've thought of every cause," Ned said as he retrieved a hematology textbook from his shelf and opened to the chapters on marrow failure. He moved his finger down a list of "differential diagnoses," the diverse categories of diseases that can deplete the marrow. Frank Hochman had considered each. We then checked the differential diagnosis of scar tissue in the marrow. Again, each cause had been addressed in the evaluation in New York, and none identified.

I walked slowly back from Ned Waterman's office to the hematology clinic, feeling unsettled. The diagnostic thinking had been complete, but there was a critical difference of opinion, based not on empirical data but on a subjective interpretation of the same marrow biopsy.

Most laypeople imagine a pathologic diagnosis to be objective and definitive. But often it is not. Very competent pathologists can view the same specimen and arrive at different conclusions.

I first would tell Alex Orkin that I had no reason to doubt the test results from New York. But as a scientist, he knew that no test performs perfectly, without false negative or false positive outcomes; in the performance of every test, there is an opportunity for human and machine error. Furthermore, diseases in the marrow are not always uniform, so there can be sampling errors. If the trocar happened to penetrate a part of his marrow cavity where abnormal cells were not growing, the diagnosis might be missed. So the flow studies and biopsy should be redone.

But I knew we had to resolve the disagreement quickly and anticipated that even after repeating these tests, we would be no closer to deciphering Alex Orkin's condition. Ned Waterman believed that the marrow cells had no intrinsic defect; the problem was external. Only special experiments might support this contention.

Such experiments on an individual patient are not undertaken lightly. They are done in a research laboratory like mine and are laborious and expensive. The scientists have to divert their time and attention from ongoing work. And the experiments are just that— not classical established diagnostic testing but new techniques to generate new knowledge. And such experiments are more prone to error than standard clinical tests like a marrow biopsy or flow studies, so they can be clinically misleading.

"I've reviewed your marrow biopsy with Dr. Waterman," I began. "It is marrow failure—meaning impaired production of blood. But we don't believe it shows myelodysplasia *or* aplastic anemia."

"Believe?" Alex Orkin shot back. "Or think? Or know?"

His deep-set eyes narrowed.

"Diagnostic tests like the bone marrow biopsy are inexact, open to varying interpretation."

"How inexact? Is the test open to interpretation on the order of 1 percent or 10 percent or 50 percent of the time?"

Alex Orkin was a physicist. He solved problems using sharp, quantitative information. But here numbers were hard to come by. Replying that medicine was not an exact science like math or theoretical physics, while true, sounded simplistic and patronizing.

"It is difficult to state just how inexact the marrow biopsy is in distinguishing among marrow failure conditions. So we look to complementary tests to refine the diagnosis. These include flow studies and DNA analysis of your marrow cells. These couldn't be done because there was no aspirate—a dry tap. We should try again and repeat the aspirate."

Alex Orkin nodded affirmatively.

"The biology of bone marrow growth is a major focus of study in my laboratory," I continued. "We experiment with marrow stem cells and try to understand what controls their maturation. I think it's worth experimenting on your marrow cells, if we can get enough. It's a long shot, but it might give us insights we don't have from standard tests."

"I like that, an experiment," Alex Orkin replied, his face brightening for the first time during our visit.

I cautioned him not to hold out too much hope for what I was proposing. The research assays were even less exact than the clinical tests that so far failed to pinpoint his diagnosis.

I then addressed an issue that had been bothering me since we first met. "Even if we come up with one of Dr. Hochman's diagnoses," I said firmly, "I disagree with the statement that you have six months to live. No one can say that, because of the inherent variability in how diseases behave."

"I'm not looking for you to sugarcoat this, Doctor. I'm not one for illusion or false hope."

"It's not false hope," I answered firmly. "Think of it in mathematical terms: There is a distribution of outcomes of any illness. Most cluster around some median and mean, but there is always variability, even in the most severe diseases. I've seen patients with brain tumors live more than two years, although the majority live six to nine months. Sure, living two years is several standard deviations from the mean, some small percentage of cases. *But it's not zero.* What's false is to say you have only some arbitrary time to live."

I saw the struggle play over his face: a theoretical physicist, trained to view variability and inexactness as enemies, trying to welcome them as allies.

"It does make a difference to hear that," Alex finally allowed. "But I still can't put much stock in chance. Now, explain the design of the experiment that you are proposing in your laboratory."

"There are three hypotheses that we'll examine," I began. "One, that some external toxin has injured or poisoned your marrow stem cells. Two, that you are 'allergic' to your marrow, meaning you made an antibody against your marrow cells and this antibody is attacking them, aborting their growth. And three, that instead of an antibody, you have white cells called lymphocytes attacking your marrow cells and destroying them."

Alex concentrated hard on my words. Ordinarily, digesting three hypotheses would not be a challenge for a person of his background and intellect. But illness and fear cloud even the keenest of minds. I took a fresh tablet of paper from my desk and sketched the scenario.

I first drew a cartoon of a bone, and then inside it a series of small circles enmeshed in a wiry matrix, like eggs in a nest.

"You'll forgive the art, but it was never my strong suit. These circles are the stem cells, the primitive marrow cells that grow and mature into all the cells of the blood. They're remarkable biologically."

I drew arrows from the stem cells and wrote the words "red cells," "white cells," and "platelets."

"The red cells, as you know, carry oxygen. Without enough,

you're anemic. The white cells defend against infection. Some do this by producing antibodies that glom on to bacteria, others by engulfing and digesting the antibody-coated bacteria. The platelets help form a clot, so without them, there is risk of bleeding."

Alex Orkin nodded. He already knew about stem cells, how they form blood, and what blood does.

"If we are successful in extracting some of the remaining stem cells from your marrow and culturing them in the lab, we can get a rough sense of their capacity to grow and mature. If they suffered a direct hit, like in myelodysplasia or some types of aplastic anemia, so they're permanently damaged, then they will fail to grow. But if you've become allergic to your own marrow cells and developed an antibody or a group of lymphocytes that is blocking their growth, then we might be able to figure that out. How?"

I left the question unanswered for a moment.

"Because when we extract the marrow, we free the stem cells from the surrounding environment, from any attacking antibodies or lymphocytes. The stem cells have a reprieve in the culture dish and can grow unmolested. Then we'll add back your antibodies to some cultures of growing stem cells, and we'll add back your lymphocytes to other cultures of stem cells. That way, we might recapitulate in the culture dish what's happening in your marrow and identify what may be blocking its production of blood."

Alex Orkin said he was following my thinking so far.

"I'm also going to perform a second experiment," I added. "I'll take normal marrow cells from a healthy volunteer and mix them with your antibodies, and, in other cultures, with your lymphocytes. I'll also take the healthy volunteer's antibodies and lymphocytes and mix them with your marrow cells—if we can get enough marrow from you. This way, we obtain more evidence that an antibody or lymphocytes in your system can inhibit marrow stem cells."

Alex Orkin smiled. A checkerboard of possibilities, he said, with two variables in play. One was his antibodies and the other his lymphocytes. These two variables would be assessed for their effects on his stem cells compared to their effects on a normal person's stem cells. This design derived from the assumption that his stem cells were intrinsically healthy, and would grow normally once removed from the inhibitory antibody or inhibitory lymphocytes.

I confirmed his succinct summary.

"Now, how might growing my marrow show your interpretation to be incorrect—and Dr. Hochman's to be right?" he asked.

"That would be suggested by your isolated marrow stem cells—freed in the culture from any external force, whether it be antibody or lymphocytes—failing to grow well."

I cautioned that such a result could still indicate an antibody at work, one whose effects could not be diluted by the isolation procedure.

"So it's inexact," he concluded. "Like most of medicine."

I agreed.

I gave Alex Orkin documents to sign. Any human specimen used for research purposes rather than standard clinical tests requires informed consent. The papers outlined a formal protocol, approved for scientific and ethical merit by the hospital that allowed me to experiment. Alex Orkin read the text. He affirmed he understood the risks of having blood drawn and marrow biopsies done for strictly research purposes: pain, bleeding, and that the experiments might be worthless.

"How long does all this take?" he asked.

"Cultures of marrow usually are assessed at twenty-one days. So, conservatively, we would have data in less than a month."

"What do we do in the meantime?"

"Continue the transfusions for your anemia and your low platelets. Your white count is low, and you're at risk for major infection. I would begin injections of G-CSF, a protein that might boost your white blood cell count."

"That's it?"

"For now, yes."

"And the marrow donor search?"

I said I'd continue searching, in case that proved to be the best option.

"That's directly opposite to Dr. Hochman's recommendation. He has me scheduled for chemotherapy next Monday and then an unmatched transplant."

"A third opinion might be helpful," I offered.

Alex Orkin shook his head no. Both Dr. Hochman and I were experts. There was nothing concrete that a third specialist would add, just another subjective interpretation of the same data.

I said I would show the biopsy to my colleagues and solicit their input as a so-called curbside consult. Then I'd call Dr. Hochman and discuss my thoughts with him.

I hoped there might be a meeting of the minds, that Frank Hochman would apprise me of something I missed that formed his opinion, or vice versa.

"Ready for me to repeat the marrow biopsy?" I asked.

Alex Orkin said he was as ready as ever.

Alex Orkin lay on his abdomen. His head, turned to the side, rested on a thin pillow. He knew what to expect from the first time. Nonetheless, at each step of the marrow biopsy I informed him of what was next.

"First the iodine."

I made a series of expanding concentric circles on the posterior crest of his pelvis using a gauze pad soaked in the brown sterilizing solution.

"Now comes the alcohol—it will feel cool."

I swabbed off the iodine with the alcohol and waited a moment for the fumes to dissipate. I ran my gloved index finger over the aseptic area to locate the prominence of the bone closest to the skin. This would be the entry point for the large-bore trocar.

"I'm going to anesthetize the area with lidocaine, like at the dentist. It's a bee sting."

The needle was 24 gauge, very thin, and drew only a single drop of blood. I raised a small mound under the skin and then rubbed it so the anesthetic was dispersed into the tissue. I then inserted the needle more deeply, all the while releasing anesthetic, until it hit the fibrous sheath covering the pelvis. I injected a large amount of lidocaine into the sheath.

"Let's wait a few minutes for the anesthetic to work," I said. "You holding up okay?"

"Fine."

"Tell me more about what you're currently working on."

"Are you familiar with superconductivity?"

I said only vaguely.

Alex explained in the simplest of terms that he was part of a team of physicists trying to increase the speed of computing. Some experts in the field believed that the maximum speeds had been defined, but he wasn't certain. His effort was on the theoretical side, creating models based on the flow of electrons in a variety of microchip materials and designs that then were formatted by others in the lab and empirically tested.

"Any successes?" I asked.

"Some encouraging results but still preliminary. I think we're on the right track, but it needs to be significantly better to be competitive, and we're far from that."

He explained he would prefer to work in purely theoretical physics, continuing his doctoral work on the behavior of subatomic particles. But he had to compromise: superconductivity was "hot" and he had to "follow the money" to be sure of support.

I took the trocar and placed a thin metal rod inside it. The rod would displace the connective and fatty tissue as the knifelike tip of the trocar cut from the skin to the bone. Once the trocar penetrated the bone and was deep in the cavity, I would remove the rod and attach a syringe to try to aspirate the gelatinous inner marrow.

I pressed downward in a swift, fluid move. The beveled edge of the trocar sliced straight through the flesh to the sheath of the bone. Alex Orkin did not move or cry out. The anesthetic seemed sufficient.

"Still okay?"

He said he was.

"Women say this next sensation is like menstrual cramps, a deep ache in the pelvis, when I bore the trocar in and aspirate out the marrow."

I pushed hard and felt the snap as the thick bone gave. I pushed further, feeling no resistance, sure then I was inside. I removed the inner rod of the trocar and attached a large syringe to the handle. I pulled on its plunger.

A thin trickle of viscous marrow slowly crept into the bottom of the syringe. I let the syringe's plunger down halfway, and then pulled forcefully upward again, maximizing the vacuum. Fragments of the inner lattice of the bone marrow, called spicules, flew up into the syringe.

"We've got some good marrow!" I exclaimed as the spicules entered the syringe.

"I felt that," Alex Orkin said thinly.

"Sorry. It's impossible to anesthetize inside the bone. But we should have cells to study."

I reinserted the thin metal rod into the trocar and pushed still deeper into the cavity. Several attempts to obtain more marrow spicules failed.

Beads of perspiration trickled down Alex Orkin's face onto the pillow. I wiped them off with a tissue.

"We're almost done," I assured him. "Just the biopsy and then the trocar comes out."

"I now have more sympathy for menstrual cramps," he replied.

I removed the inner rod and twisted the open trocar, feeling the inner latticework grate along the sharp edge of the instrument. I twisted again, forcefully, and then quickly pulled the trocar out. Blood welled from the hole the instrument made in Alex Orkin's skin. I applied a pressure bandage at the bleeding entry point.

"Take slow, deep breaths," I told Alex Orkin, the sweat now streaming down his cheeks.

I teased out a core of marrow from inside the trocar. It looked like a thin calcified worm. This was the biopsy. Part of it would be placed in fixative, and the rest put in sterile saline for special tests to be done by Ned Waterman.

Alex Orkin was still breathing hard. The biopsy always hurt, regardless of the physician's skill.

"I've had eight marrows done on myself," I informed him.

He turned his clammy face to me.

"Into S and M, Doctor?" he said with a thin smile.

"No." I grinned back. "But when I was a research fellow in the lab, we would aspirate marrow for experiments from each other. Since my back surgery, I don't participate. Now we pay volunteers."

"How much do they get?"

The going rate was seventy-five dollars.

"Times must be very hard in Boston," he said with a widening grin. "You couldn't get away so cheap in New York."

The precious marrow aspirate was divided into two parts: one went to Ned Waterman for standard analysis, the other to my laboratory for experiments.

Three days later Ned and I again sat at the microscope. After careful inspection, our conclusion was the same: it was marrow failure, but didn't look like myelodysplasia or aplastic anemia. The

DNA studies on the aspirated marrow cells would take more time. The flow studies again were indeterminate. I called New York.

"All is well, Jerry?" Frank Hochman asked.

I said things were.

"Planning to attend the stem cell meeting at Keystone? If so, I'd like to take you down the mountain with me," Frank continued with a chuckle.

The year before, at a winter hematology research meeting held at a Colorado ski resort, Frank and I had shared a chairlift. When we reached the top, he challenged me to follow him down his favorite double–black diamond expert trail. It was far beyond my ability, and I opted for a less harrowing, intermediate-level run.

"I'll pass again and keep living as a coward, with intact limbs," I replied.

I then presented Ned Waterman's assessment of Alex Orkin's biopsy and our opinion of the case.

"We'll have to agree to disagree," Frank said with polite firmness. "The senior man down here, Paul Zilber, is as accomplished a pathologist as any. I've followed this patient now for nearly two months. The longer we delay, the riskier it gets. Professor Orkin needs definitive treatment, and he needs it now."

I asked if there were any progress on identifying compatible donors.

"We're going ahead with an unmatched transplant," Frank Hochman answered.

My stomach tightened. The transplant was more dangerous than charging down a double–black diamond run. It was falling off a sheer cliff.

If Alex Orkin didn't die from the procedure, he would likely live debilitated with chronic graft-versus-host disease.

"What about a trial of growth factors to boost his blood counts, as a temporizing measure?" I offered. G-CSF, the white cell growth factor, and erythropoietin, the red cell growth factor, might increase his marrow production.

"If it's aplastic anemia, they don't work," Frank replied, his voice moving to a sharper register. "And if it's myelodysplasia, G-CSF may spark full-blown leukemia. No—this is no time for 'temporizing measures.'

"I know the risks here," he continued. "An unmatched transplant might succeed one out of four or five times—or less. And he'll have some form of chronic graft-versus-host disease if he survives. But I've seen countless cases of severe marrow failure in my thirty years of practice. It's a bad disease—a fatal disease. As I like to tell my trainees: 'Desperate diseases require desperate measures.'"

Don't just do something—stand there! This injunction broke in my mind like a crashing wave. *Don't just do something—stand there!*

I first heard this dictum as a medical student. I was on the wards at Columbia Presbyterian Hospital with Dr. Linda Lewis. A tall, statuesque woman raised in the no-nonsense culture of rural West Virginia, she was one of the most respected clinicians at the hospital. One day on rounds I presented a complex case to her. She repeated my physical examination and then suggested we retire to a conference room to discuss the patient. I confessed that I wasn't sure of the diagnosis. Dr. Lewis said plainly she wasn't sure either and couldn't think of other tests likely to clarify the case. In situations like this, she asserted, it can be best that you "Don't just do something—stand there."

Her recommended restraint felt deeply unsatisfying, and I chafed at it. It was counter to the core of modern clinical training. We were bred with the tenet that it was a sacrilege not to use powerful technologies at hand. Moreover, we understood that needy patients wanted us to act, and we were told that our acts, in and of themselves, had therapeutic impact. "Proactive" was the laudatory term. A passive doctor revealed his ignorance and risked undermining his credibility in the eyes of patients and their families. But throughout my clinical life, the contrarian West Virginia wisdom of Linda Lewis stood me in good stead.

I also came to see that the mark of a talented and confident clinician, like Dr. Lewis, was the ability to say "I don't know." It revealed a humility and honesty, and a solid confidence, that I aspired to.

"I'm going to advise Professor Orkin to wait," I said haltingly.

There was cold silence on the line.

I was on shaky ground. I had no solid evidence against Frank Hochman's position. But the more I pondered his advice, the more I sensed it was wrong. The risk and degree of harm were simply too great.

"I'm seeing the patient today," Frank Hochman retorted with unmasked ire. "And I'll convince him to do what's needed."

Alex Orkin telephoned me later that afternoon. He peppered me with questions: How can there be such discord about the same biopsy? What would I estimate as the probability that I'm wrong in my analysis? One percent, 10 percent, 50 percent? What if the experiments in my laboratory were indeterminate in outcome—would that change my opinion? Or if they suggested an intrinsic problem in the marrow stem cells—would I then support Dr. Hochman that he proceed with an unmatched transplant?

Sandra, Alex's wife, was also on the call. She mainly listened, only occasionally adding a few words to extend her husband's questions.

I answered each query as specifically as I could. But, in essence, there was one reply to them all: intuition can weigh as heavily as raw data in clinical judgment, and competent consultants can sharply disagree, because of their own perceptions of the risks and benefits of an intervention.

"Those are not very reassuring statements," Sandra Orkin said.

They were not. I was not in a position to be reassuring, I admitted, only very frank.

"But we have to go one way or the other," Alex Orkin said.

I again suggested a third opinion.

"I don't see this as a tiebreaker vote," Alex Orkin asserted. "And as you imply, the third opinion will depend on the 'perception' of that consultant, so will be subjective, not objective. And Dr. Hochman would direct me to someone of his philosophy, and you might refer me to someone with yours."

I told the Orkins that Ned Waterman was going to send the marrow slides to other pathologists.

"That could take weeks. And Dr. Hochman told us we had no time to wait," Sandra Orkin said.

"But, honey," Alex Orkin interjected, "he also said, again, that I had less than six months to live."

I remained silent.

"And today, when I asked him how he calculated that—specifically, in my case—he answered, 'I know'—that he was the professor of hematology, and patients like me who keep questioning his every statement can see his curriculum vitae."

I still did not speak.

"I don't like statements like that," Alex continued. "They're meaningless, in science or in any other rational discipline. I am not one to accept his opinion as a fiat based on his credentials or reputation. So I pushed him further, asking for a set of confidence limits around his estimate of my survival. He told me I was being 'tangential' and I was 'in denial.'

"I told Dr. Hochman I am painfully aware I have a life-threatening condition, and I've never had a delusion about that—or anything else—in my life."

Alex paused.

"But then again, I don't fully trust my thinking. I feel as if I'm looking at this whole situation from the outside—observing from different positions and seeing events move parallel to me."

"That's a common feeling," I said. "But it's not denial. It's just hard to grasp the reality of sickness when it occurs so quickly, out of the blue. It seems surreal."

"It's not surreal to me," Sandra Orkin said plaintively. "Alex, we have to decide."

I waited nearly a week for the Orkins' answer. I could not move the case from the front of my mind. I reread my notes and Frank Hochman's records. I kept trawling over Alex's medical history and tests, looking for the nth time for some overlooked clue. I found nothing.

The call came late Friday morning. It was Sandra.

"He's in the ICU," she said, her voice trembling.

I listened with tense anticipation, my stomach churning.

"I brought him to the hospital just before midnight. He was delirious—ranting—I've never seen Alex confused before. It was awful."

Alex Orkin had a fever of 104 degrees. A pneumonia was diagnosed, she said, antibiotics started, and oxygen given by a face mask.

"If he doesn't improve quickly, Dr. Hochman thinks he'll need a respirator."

I swallowed hard. Alex Orkin's white blood cell count was very low, his immune defenses minimal. Death from infection in such settings was frequent. Perhaps I was the one who was in denial, heeding some cowardly inner voice saying "Wait, wait."

Sandra said she would be in the visitors' area of the ICU if I wanted to contact her. Her mother, who lived near the Orkins, was watching the twins.

I reached Frank Hochman around noon. His voice had the hoarseness of a sleepless night.

"The blood cultures are already positive for gram positive cocci," he informed me. "The sepsis is high grade. We've got him on triple antibiotics."

I had little to say, except to ask to be kept informed.

"You call me," Frank Hochman replied acidly. "My nurse is available to update you as needed." He paused. "If we pull him through this, I'm going to push the Orkins hard to go immediately to transplant. It's madness to just sit and wait for the next catastrophe."

⁓

Alex Orkin was placed on a respirator at the end of that first day in the ICU. The pneumonia had blossomed in multiple lobes of his lung, and he wasn't able to sustain a satisfactory level of oxygen. Over the ensuing week, his red blood cell and platelet counts plummeted—typical with systemic infection where blood cell survival is shortened. Multiple transfusions were given but hardly kept pace.

I called Dr. Hochman's office twice a day. I spoke with his nurse, a pleasant older woman who had a real-time grasp on Alex's status. Eight days into the hospitalization, Frank Hochman paged me.

"We may lose him," he said grimly. "We can't keep on top of the pneumonia."

"You're between a rock and a hard place," I replied, reverting to clichés as often is the style in discussion of complex clinical cases. "Should we try to boost his white cells with G-CSF? It may not work if it's aplastic anemia, and—"

"And trigger leukemia if it's myelodysplasia," Frank Hochman finished my sentence.

"True. But the risk of triggering leukemia seems smaller than the risk of no white cells."

Frank Hochman was silent.

I waited for him to speak, thinking that the risk of sparking leukemia was theoretical. Sporadic cases were reported in myelodysplasia, but it was hard to know whether it was the G-CSF or just the natural evolution of the disease into acute leukemia. I quietly wondered if Frank would be willing to change his mind. When I had attended his presentations at hematology meetings, he bristled at pointed questions. And the bitterness in instructing me to call his nurse, and only now contacting me when the situation was so desperate, revealed a dangerous streak of pride.

"There's really little we can lose at this point," I added.

"Okay. I'll start it. At a high dose. If it has any chance of working, it'll be at a high dose. But if he develops acute leukemia, it's on your head."

Later that day I opened the incubator in my lab and retrieved a stack of petri dishes.

It was only fifteen days since we aspirated the marrow from Alex Orkin and obtained a comparative specimen from a healthy volunteer. But I was desperate for any bit of data on his disease.

Each petri dish was filled with agar, a gelatinous material that resembles straw-colored Jell-O. I examined the dishes "blind," meaning I didn't know beforehand which were holding Alex's cells and which the normal volunteer's. The plates were given coded numbers as a further safeguard.

The sharp light of the microscope diffused through the thick agar, casting a hazy glow like the corona of the moon and illuminating clusters of translucent marrow cells. I studied some twenty dishes, systematically counting the numbers of clustered cells by quadrants. I jotted these numbers in columns on a sheet of lab paper.

With me was Yigong Chen, a researcher in my lab expert in the biology of marrow growth. He independently counted each dish.

"No difference," I concluded.

Chen nodded his agreement.

"You're certain your technician seeded each dish with the same numbers of cells from the patient and the volunteer?"

He opened his lab book and showed me the experiment. It was done correctly.

I returned the dishes to the incubator and then removed a second stack. These again held marrow cells cultured in the agar matrix, but here Chen's technician had added different amounts of Alex Orkin's antibodies or the healthy volunteer's antibodies. Again we counted the cultures blind. There were roughly the same numbers between Alex Orkin and the healthy volunteer.

I counted the third set of petri dishes, and then Chen did. In these, his technician had mixed the lymphocytes from Alex Orkin's blood, or the lymphocytes from the volunteer, with the marrow cells.

He finished and excitedly turned to me. We both found that some of the plates were nearly barren of cell growth.

"Break the code now," I said. "I'm not waiting a full twenty-one days."

In this third set of cultures, where Alex Orkin's lymphocytes were mixed with his own marrow cells, or with the volunteer's marrow, there was sparse growth in the petri dish. But the volunteer's lymphocytes had no effect on Alex's or the volunteer's marrow.

We still did not have a diagnosis but at least had a lead. There were lymphocytes in Alex Orkin's system that inhibited the growth of both his marrow and the marrow of the normal woman volunteer.

I held back from embracing these results wholeheartedly. There was an important caveat. Alex Orkin had been transfused many times. It was possible that this exposure to foreign blood sparked an immune response among his circulating lymphocytes, so that they inhibited the volunteer's "foreign" marrow. That, of course, did not explain why Alex Orkin's own marrow would be suppressed by his own lymphocytes. Nor did that explain the robust growth of his marrow in the absence of these added lymphocytes. These last results suggested that his stem cells might grow under the right conditions.

"The lab tests help in distinguishing what it isn't," I explained by phone to Sandra, trying to express the data and the caveat as clearly as I could. "Better said, there doesn't seem to be a major intrinsic abnormality in Alex's marrow cells. They can grow if given the right environment."

"That gives me some hope," she replied in a wan voice. "I just keep thinking he's going to die, and I'll be alone with the twins."

"There is hope," I said, struggling to sound confident. "There's always hope."

For seven harrowing days Alex Orkin's pneumonia raged. His blood oxygen was difficult to sustain even with maximal pressure from the respirator. Then, on the eighth day, the G-CSF began to take effect. The white count slowly climbed to the cusp needed to combat bacteria. Over the ensuing week, as the white cells rose, the infection started to abate. Less pressure from the respirator was needed. The chest X ray started to improve. Alex was gradually weaned off the respirator. After a full month in hospital, he had lost nearly thirty pounds and could not walk more than a few steps unassisted. He was discharged home. There was no sign of acute leukemia.

"I'd continue the G-CSF," I said to Frank Hochman.

"Perhaps," he replied dryly. "We've been lucky so far. I'm concerned that long-term treatment will exhaust his marrow."

Frank's objection was that chronic G-CSF therapy, relentlessly driving the fragile marrow stem cells to grow and mature into white blood cells, would somehow deplete them. This theory was widely aired when growth factors like G-CSF first came into clinical use, and still loomed as a specter in the background, but there was little evidence that such "marrow fatigue" occurred.

"I feel very strongly about continuing the G-CSF," I asserted, "given what he just went through, and the fact that it had some effect."

"We found several closer donor matches," Frank Hochman said, changing the subject.

"How close?"

"Three out of six."

This meant that half of the genetic determinants were in accord. Transplant was still very high risk.

"I'd give him more time," I said. "I'm intrigued that his marrow cells grew so well in the lab, and that his counts improved on the G-CSF."

"It's awfully thin to hang your hat on that. Honestly, I think you're chasing rainbows."

———

I telephoned Alex Orkin later that day. "It must feel good to finally be home. I'm sorry you've had such a difficult past few weeks."

"I don't want sympathy—I want answers," he shot back, the rattle in his voice audible after the pneumonia.

"That's what we're working toward—answers. It just takes more time," I replied gently.

"How long?" Sandra cut in. "Days, weeks, months? Alex's pneumonia was a warning, Dr. Hochman said. He says now is the time to go ahead with the transplant."

"But there is still no matched donor," I cautioned.

"He says to just do it, with a partial match, three out of six."

I paused to collect my thoughts. Alex had faced death and suffered greatly, Sandra witnessing his travail daily at his side. Now a powerful current of panic was sweeping them forward—panic being perhaps the greatest danger to clinical decision making, drowning both reason and intuition in its wake.

"Despite how grim things look, there is no benefit in panicking," I calmly said. "Let's step back and look at the data that are encouraging."

I listed them: the robust growth of his isolated marrow cells in my laboratory; the rise in his white cells on G-CSF; and, recently, the report of no DNA abnormalities in his aspirated marrow cells.

"They're largely circumstantial data," Alex said.

"True, but they're consistent."

"Honestly, we just don't know what to do, whom to believe," Sandra said in a quavering voice.

"Give yourselves more time to think. Not weeks but a few days. Outline your thoughts, pro and con. Generate questions, and call me back to clarify or expand on them."

I was trying to substitute thought for fear, and to provide a structure to reason. They might still opt for the transplant, but it would be a sober step.

"The decision, of course, is ultimately yours," I added. "And if you choose to follow Dr. Hochman's advice, I'll still help in any way I can."

I heard back from the Orkins later that week.

"Dr. Hochman says I'm 'diddling around,' " Alex reported.

I listened silently.

"And he said there has to be one chef in the kitchen—that I'm a Ping-Pong ball, bouncing between you and him."

I replied that Hochman was right. Ultimately, there has to be one physician directing a case.

"But I didn't like his reply when I asked again for his thinking about the wisdom of a transplant. He said that his judgment was superior to yours—that he has more experience with transplantation than you. When I reminded him he had opposed the G-CSF, he answered that he still felt it was risky, that it might exhaust my marrow."

I answered that that was a theoretical consideration, and the benefit far outweighed the risk.

Of course, I was hearing what Alex Orkin had heard, not exactly what Frank Hochman had said. But it held the kernel of truth. Frank was more experienced with transplantation than I—he headed the program at his hospital. But that, paradoxically, could be interpreted as a caution, since we all tend to fall back on what we know best when giving advice. As one colleague put it: "If the tool in your hand is a hammer, almost anything can look like a nail."

"I'm going to wait on the transplant," Alex Orkin stated, "and follow your lead. You are the chef in the kitchen. At least for now."

"He's going to die, you know," Frank Hochman said gruffly. He was bitter about the Orkins' decision. "They said they still wanted to see me, to have his care through my office. But I find it difficult to attend to someone who rejects my advice."

I replied that Alex Orkin would benefit from Frank's expertise, that despite the choice to wait, there was an established relationship. Finding a new doctor in New York would be logistically disruptive. Moreover, transplantation might emerge as the best option if a match were found.

"Let's agree to disagree for now," I offered, returning Frank's earlier phrase, "and continue to assess the situation."

We restarted the G-CSF to try to boost Alex Orkin's white cell production, and added another blood cell growth factor, erythropoietin, to try to do the same for his red cells. I was tempted to give immune suppressive drugs again: if his lymphocytes were suppressing his marrow growth, then immune suppressive therapy might block them. But I refrained. Immune suppressive therapy was a shotgun treatment, blocking beneficial as well as aberrant cells, and could make him susceptible to infection. The last thing we needed was another siege.

"I understand your thinking," Alex Orkin replied. "You don't want to change too many variables at one time in case there is an effect— positive or negative. If my marrow production improves, it could be due to the G-CSF and erythropoietin and/or the immune suppressive drugs. If my marrow production worsens, it could be due again to the growth factors causing marrow exhaustion and/or the immune suppressive drugs reactivating some microbe. I'm game to go a step at a time."

Each Monday and Thursday, I received a fax from Alex. It summarized his current symptoms and his cumulative doses of G-CSF and erythropoietin, and included an updated graph of his blood counts. They were inching upward. He was still fatigued, napping in the afternoons, but felt more energy overall. He had gained four pounds but was forcing himself to eat, his appetite still poor. I received regular reports from Frank Hochman's office as well, copies of the blood tests and physical examinations.

"I'd like to repeat the bone marrow," I said to Alex Orkin, "just after Labor Day."

Alex said he had taken a medical leave for the spring semester, and classes began after the holiday. He was anxious to return to work.

"The boys also start preschool, and I want to be there with Sandra for their first day."

There was no rush, I said. We set the appointment for the following month.

It was a cool day in early October, a brisk wind testing the hold of the leaves as they began their autumnal change. The Harvard Medical area was dense with traffic and pedestrians, classes in full swing after the summer break, hospital staff all returned from vacation.

Alex Orkin sat in the first row of chairs in the clinic waiting area. He was reading a document. A thick sheaf of papers was on his lap. He didn't look up until I greeted him by name.

"Oh, sorry," he said warmly. "It's a new grant proposal I'm preparing."

He explained that it was on neutrinos. He wanted to move away from superconductivity and do what he really wanted to. It might take several submissions for the application to be funded—if it all. But it was worth the effort.

Alex undressed. His color was good, without the earlier anemic pallor. He had gained another six pounds. There were some harsh breath sounds, his bronchi not fully healed, but no other abnormal findings on physical examination.

"I haven't been looking forward to this. Did you really do marrows on each other in your lab?"

I assured him we did. And that we weren't into S and M.

This time it was easy to obtain an aspirate. The thick juice flowed up into the syringe like bubbling crude from a brimming well.

"Any inflation in the price of healthy volunteer marrow?" Alex Orkin asked as he dressed after the procedure.

"No. This is Boston, very conservative and fixed in its ways."

"I feel like I should contribute in some way," he said more seriously. "Like pay for the experiments you did."

I appreciated the thought but said it wasn't necessary. We had charitable funds in the lab for such work.

Three days later I sat in the pathology department.

"Remarkable," Ned Waterman said, surveying the specimen under low power and then high magnification. "Truly remarkable. I've never seen such a dramatic change."

A pulse of electric joy ran through me. For months I had anticipated this moment, but had not permitted myself to overreach.

The wide desolate tracts that once marked his marrow were now filled with streams of developing blood cells.

"There's even little scar left," Ned added. He paused thoughtfully. "Will you test the miracle by cutting back on the G-CSF and erythropoietin and seeing what happens?"

I hesitated before answering. It would be *interesting* to know whether these growth factors were still needed, or if blood could be produced without them. Stopping then would be a clinical experiment. On the other hand, there were clear risks involved. The marrow had not fully recovered in terms of blood cell production: Alex Orkin was still somewhat anemic and his platelet count, while in a safer range, not quite normal. Furthermore, not identifying what caused his marrow failure meant that the mysterious factor or disease could still be lurking. The marrow might grind to a halt again after stopping the G-CSF and erythropoietin. And there was no confidence that reinstituting them would work again—that, too, would be a clinical experiment. My curiosity and scientific nature yielded to another maxim of Dr. Lewis: "If things are going well, continue what you're doing."

I left Ned Waterman and quickly called the Orkins with the news. Sandra began to cry, and I felt my throat tighten.

"How did you know the right thing to do?" Alex Orkin asked, his voice high with elation.

"I was guided by a dictum that I learned in medical school from a very wise and experienced senior physician. When dealing with a case where the diagnosis was not clear and treatment very risky, she cautioned me to 'Don't just do something—stand there!'"

"But 'just standing there' might have proved more dangerous than the transplant," Sandra said. "Alex almost died from the pneumonia. He was a setup for another disaster."

"My intuition could well have been wrong," I admitted. "Maybe we were just lucky, a coin toss."

"You can see it as luck, as a coin toss," Alex Orkin stated, "but I don't. Think like a physicist for a moment," he continued. "What you call luck just means a positive outcome and lack of a full data set to know why. Take that coin toss: if I knew the position of the coin, the force used to flip it, the frictional drag of your fingers and the air, the height it reached, I could calculate its landing as heads or tails. In my case, there were several discrete bits of data that you linked together in a linear fashion. And what people call intuition is of course more than that. There is a rational basis for it. It's just that you didn't have a full data set, and your reasoning was below the level of your awareness."

Perhaps, I silently said to myself.

I still was not completely sanguine about his clinical status. Alex may have had a reaction to some unknown virus. This could account for his period of fatigue prior to the faculty Christmas party. The mystery virus suppressed his marrow by turning on the cadre of lymphocytes detected in my experiments. Over the course of the past year, the reaction to the virus spontaneously subsided. And if this were true, the virus might still be in his system, dormant, to one day reawaken and wreak havoc with his blood. Another scenario was that the cadre of lymphocytes was not sparked by an undetected virus but represented the vanguard of an autoimmune disease, like lupus. Dr. Hochman's initial immunosuppressive therapy had serendipitously contained the disease before it was fully manifest. It, too, could return.

But this was all speculation. I knew I would follow Alex Orkin with a restless vigilance, all the while hoping that hematology would advance, and better tools would be created to characterize syndromes of marrow failure and guide rational therapy. We had stored his antibodies, lymphocytes, and stem cells in my laboratory. One day we might retrieve them, and using those new tools, solve the mystery of his disease. We would understand why it began and why it passed, and how to treat it. We would have a full "data set," as Alex put it. But even then, even with all the clinical information, medicine was not physics. The inherent variability of human biology meant judgment could not be reduced to mathematical calculation. Intuition, and luck, would still count.

Epilogue:
Second Opinions

I recently saw a middle-aged man seeking a second opinion. He had had months of unexplained fevers and underwent exploratory surgery. Both his spleen and abdominal lymph nodes were found to harbor two distinct types of lymphoma. "It's lightning striking twice," the referring oncologist said when he called me about the case. The medical literature had no precedent to guide us, so clinical decisions had to be made intuitively, drawing on knowledge about each type of lymphoma. The patient visited yet a third expert, then returned to his primary oncologist. A consensus was reached among us, and I continued to confer through the course of the empiric individualized therapy. A complete remission of both cancers was achieved.

Second and third opinions are customary in cases like this one, when the illness is rare and treatment is unclear. Similarly, second opinions are usual for diseases that are not rare but are life-threatening. Here, available treatments usually have a high risk of debility or even death. Experimental therapies are frequently considered for life-threatening disorders, and an assessment of their side effects and rationale is vital before entering the clinical trial. This is best done by conferring with an expert who is not himself invested in the testing of the drug.

Fortunately, most medical decisions do not require second opinions. Our regular care is readily and efficiently accomplished by the primary physician. We receive antibiotics for strep throat or a sinus infection, vaccines for prevention of influenza, antihypertensive medications for elevated blood pressure not controlled by diet, and other standard interventions for a host of common problems. Yet, even when dealing with the routine, both doctor and patient must be vigilant. There is some degree of uncertainty, and thus some degree of risk, inherent in every clinical intervention. This bond of mutual vigilance between doctor and patient is forged through the melding of knowledge and intuition.

Often a patient and his family hesitate to ask if a second opinion would be beneficial because they fear they will "insult" the doctor, that the question will be misconstrued as a threat and alienate their caregiver, who might then abandon them out of pique. This should never occur. A patient and his family should never hesitate to seek more advice and counsel.

"There should be no ego involved in getting you the best care," I state explicitly when a patient or a family raises the issue of a second opinion. I am speaking to myself as much as to them. A doctor has to learn to subsume any prideful feelings. A second opinion tacitly highlights his limits and fallibility. "If it means being cared for in another center, in another physician's hands, because something can be done better, then you should be there." I know I mean it, because I would mean it for myself, my wife, my children.

There is another consequence to my patient's query about a second opinion. It causes me to wonder whether I failed to fully communicate to him my thoughts and my understanding of his condition. His question about a second opinion may be his way of saying that I need to reopen our dialogue, to listen again, more carefully, to his words.

Careful listening, I have learned, is the starting point for careful thinking. We all sense when someone is listening carefully to us, and when he is not. In the latter instance, a doctor comes with blinding preconceptions, or is distracted, or is arrogant, and cuts things short. My wife, Pam, felt this when the pediatrician in Connecticut dismissed her sense that our infant son was seriously ill. His response to her words sounded as if she had not spoken. This failure to listen put us on a path that nearly resulted in the loss of our child.

The same flaw occurred when we were in the emergency room. The surgical resident was tired and overtaxed, anxious to get some sleep, and not really paying attention to us. This realization caused us to doubt his judgment, and catalyzed our decision to stop being passive and to go around his advice. We were trying hard not to be demanding, second-guessing doctor-parents, trying hard to accommodate to the system. Ellen O'Connor, the data processor whose child was suspected of having an intussusception, interestingly was quicker to challenge the ER intern when she concluded he ignored her intuition. Like Ellen, one should, in such circumstances, forcefully assert one's right to see the attending physician, who is senior and seasoned, regardless of the time of day or the scowls of the house staff. Patients and their families feel vulnerable and uneasy challenging authority—but sometimes it must be done.

A doctor's answers to a patient and his family should make sense to them. The nature of the medical problem, what is expected and what is unpredictable about its course, the level of risk inherent in the recommended therapy, all need to be clear. There is nothing so arcane or technical in medicine that it cannot be explained to, and understood by, a layperson. If such an explanation is not offered, then there is cause for concern. The doctor may not himself fully comprehend the situation, or might not be willing to disclose the breadth of issues involved, or has failed to fathom the patient's intuition. In such instances, a patient and his family should seek a second opinion.

It takes a physician time to do this, to explain a clinical situation, and if necessary reopen a dialogue and refine his words. Time is the least available element in the new world of medicine. Managed care treats it as a tightly controlled commodity. Follow-up appointments with the doctor are often measured in single-digit minutes. This is part of the growing pressure on caregivers to economize in all aspects of their work. Rewards are given when the maximum number of patients is seen in a minimum number of hours. This factory mentality further dictates that the fewest tests be offered and that the cheapest therapies be chosen. Some clinical practices seek caregivers with little experience (who, therefore, command lower salaries) to act as primary arbiters of diagnosis and treatment. Certainly, some of these changes are a response to egregious abuses of the past fee-for-service system. But the pendulum has swung too far in the opposite direction. Managed care is designed for the standard and the generic, and functions poorly in addressing the unusual and the individual. This was the painful truth for Isabella Montero. Sadly, the tightening shackles of business on medicine particularly restrain a request for a second opinion. Most managed health plans limit consultations to specialists within their own network. Only when the demands of the patient and family are very vocal may there be loosening of such restrictions and freedom to consult a chosen doctor, or be cared for at his hospital.

Another malignant manifestation of money in medicine is seen when medical centers and physicians seek to "hold on to" patients who are well insured, famous, or possible donors. These considerations certainly affected Robert Beckwith's initial care. Most physicians are ethical and professional, and in the same way they learn to subsume their egos, they learn to subsume their natural desire for personal or institutional gain when that conflicts with what is best for their patient. Resistance to a request for a second opinion can be a red flag that such territorial agendas are at work.

There may also be other unstated agendas that mar a patient's care. When my grandfather was seen by Dr. Mathers, my mother and I were not given a detailed picture of the goals of the hospitalization. It is very appropriate for a family to ask a physician to fully describe his rationale and plans for diagnosis and treatment. If one reason is to conduct research and gather data, then that aim must be openly stated and of course pursued only if it is in concert with the patient's and family's needs. Questioning should be welcomed by a doctor: it gives him the opportunity to learn more about the patient, his family, and their wishes. Dr. Mathers was vague and evasive; in retrospect, our inquiries should have been more pointed.

Physicians certainly question each other. Sometimes, as in Alex Orkin's case, there can be a sharp difference of honest opinion about what to do. Dr. Hochman had no nefarious or hidden agendas. This is perhaps the most complex situation for a patient and his family. One guide to decision making is an ancient clinical imperative attributed to Hippocrates: "First, do no harm." A clear sense of the balance between risk and benefit for the individual patient is key in choosing between differing experts.

We seem distant from the days of Hippocrates. Information about genes and their roles in disease is unfolding at an extraordinary rate. This knowledge can outstrip the physician's ability to state a prognosis and offer validated therapy. Yet the ancient wisdom of carefully weighing risk and benefit still applies, as seen in the case of Karen Belz. Her struggle to know if she carried the BRCA mutation, and how to act on that knowledge, will be replayed many times with many other disease-causing genes in the upcoming years. Patients, their families, and physicians should be particularly open here to diverse opinions and debate, and seek the input of all credible sources in formulating choices that have no precedents.

Similarly, laboratory research is providing many new and powerful drugs, all of which must be experimentally tested in clinical trials prior to prescription use. The patient who is considering an experimental treatment needs to be objectively informed about the nature of the therapy and expectations of its benefits. While each new drug sparks enthusiasm and hope, sober realism is imperative. I learned this lesson with James Leahey.

James's story also brings into focus the element of chance in clinical medicine. I think of chance as a player in the background, a silent partner to a stated plan of diagnosis and treatment. We are tempted to ignore its existence, to proceed as if we are entirely in control. But that is a risky illusion. When we deny chance's role, ignore its presence, we become more vulnerable to its upheaval—unnerved by its appearance and therefore unable to find our way around the detour it causes. Paradoxically, when we expect the unexpected, when we accept the uncontrolled element of chance in clinical medicine, then we better control its effects. We are less unnerved, more able to maintain our footing and reset our compass.

This lesson was painfully learned when I was a patient and faced stark uncertainty. Frightened, confused, and demoralized, I found it hard to listen and even harder to hear. I felt lost and desperate, and could not summon my own intuition. I was too ready to put myself entirely in the hands of a doctor who seemed to have all the answers, and I discounted the sober perspective of my wife, Pam.

What I experienced is, to a varying degree, the experience of every patient in the grip of illness. During the months after the disastrous and unnecessary surgery, I revisited in my mind each step that had brought me to such debility, and saw in a harsh retrospective light how I had relinquished my critical judgment. It was only when I began to question and evaluate the opinions offered to me that I reclaimed my intuition, and found the physicians and caregivers who helped me to regain my health.

Evaluating medical advice is the greatest challenge for every patient. Armed with knowledge, steadied by family and friends, and calling on intuition, we can gain clarity and insight, and are prepared to make the best possible decisions.

Acknowledgments

I am fortunate to live and work in several different worlds: the world of clinical medicine and patient care; the world of science and laboratory research; the world of family and friends; and, recently, the world of writing and communication. I benefit greatly from experienced guides in each of these worlds.

My wife, Pam, is an outstanding physician, an endocrinologist specializing in disorders of metabolism. Her experienced and discerning eye is the first to see my writing. Much of what we share in attitude and aspiration forms the fabric of these stories. The dedication of this book does her only partial justice.

It was during an initial discussion with Henry Finder, my editor at *The New Yorker,* about the BRCA gene and breast cancer, that the idea for this book grew in my mind. I am deeply indebted to him, and to Dorothy Wickenden, Tina Brown, and David Remnick for refining my thinking about how to express the many dimensions of medicine and human biology.

Suzanne Gluck at ICM, and her most talented assistant, Karen Gerwin, do much more than bless an idea. At each step of its gestation, birth, and maturation, they offer a steadying hand and wise counsel. No author could ask for more.

Dawn Drzal, my initial editor at Viking Penguin, shuttled seamlessly between the macro and micro dimensions of a work-in-progress, helping sculpt theme and detail with the touch of an artisan. After Dawn left, Janet Goldstein expertly shepherded the work to completion. Phyllis Grann, Susan Petersen, Paul Slovak, and Barbara Grossman all provided vital support to the project. Jane Praeger, working with Viking, was extraordinarily thoughtful in sharpening my communication of theme and content, and in guiding the book's presentation. She is a gem.

Margo Howard, David Sanford, Arthur Cohen, Stuart Schoffmann, Sally Button White, and Ruth Bayard Smith, friends with very special skills, graciously read early drafts and helped me focus my language and thoughts. Abe and Cindy Steinberger, Youngsun Jung, Charlene Engelhard, Annette de la Renta, Elizabeth Weymouth, Ellen Murphy, Caroline Alexander, Jan Mitchell, Andrew Sullivan, Liz Young, Loretta Itri, Nancy Breslin, Roberta Ferriani, Francine Pascal, Michael and Juliana Rothchild, and Marty and Anne Peretz offered important feedback and encouragement. Certain parts of the manuscript were critiqued by Lenny Groopman and Yasmine Ergas (my brother and sister-in-law), Francine and Harry Hartzband (my in-laws), and Stephen and Georgia Nimer. A special thanks to Frank Rich, Alex Witchel, Melanie Thernstrom, and Ingrid Sischy for sage advice and warm support.

My mother, Muriel Pollet, and stepfather, Maurice Pollet, were essential in the writing of my grandfather's story. Their strength and courage during Grandpa Max's illness inspired me.

In all chapters, certain personal details and circumstances have been changed to protect confidentiality and clarity. Otherwise, the stories as told are those of my patients, my family, and myself. To my patients, I owe a degree of gratitude that cannot be captured in words.

Index

About the Author

Jerome Groopman, M.D., is the Recanati Professor at Harvard Medical School, Chief of Experimental Medicine at Beth Israel Deaconess Medical Center, and one of the world's leading researchers in cancer and AIDS. A staff writer in medicine and biology for *The New Yorker*, he contributes regularly to *The New York Times*, *The New Republic*, *The Wall Street Journal*, *The Boston Globe*, and numerous scientific journals. He lives with his family in Brookline, Massachusetts.